AF455111

CIRCUIT DESIGN FOR WIRELESS COMMUNICATIONS

THE KLUWER INTERNATIONAL SERIES IN ENGINEERING AND COMPUTER SCIENCE

ANALOG CIRCUITS AND SIGNAL PROCESSING

Consulting Editor*: Mohammed Ismail. *Ohio State University

Related Titles:

DESIGN OF LOW-PHASE CMOS FRACTIONAL-N SYNTHESIZERS
DeMuer & Steyaert
ISBN: 1-4020-7387-9

MODULAR LOW-POWER, HIGH SPEED CMOS ANALOG-TO-DIGITAL CONVERTER FOR EMBEDDED SYSTEMS
Lin, Kemna & Hosticka
ISBN: 1-4020-7380-1

DESIGN CRITERIA FOR LOW DISTORTION IN FEEDBACK OPAMP CIRCUITE
Hernes & Saether
ISBN: 1-4020-7356-9

CIRCUIT TECHNIQUES FOR LOW-VOLTAGE AND HIGH-SPEED A/D CONVERTERS
Walteri
ISBN: 1-4020-7244-9

DESIGN OF HIGH-PERFORMANCE CMOS VOLTAGE CONTROLLED OSCILLATORS
Dai and Harjani
ISBN: 1-4020-7238-4

CMOS CIRCUIT DESIGN FOR RF SENSORS
Gudnason and Bruun
ISBN: 1-4020-7127-2

ARCHITECTURES FOR RF FREQUENCY SYNTHESIZERS
Vaucher
ISBN: 1-4020-7120-5

THE PIEZOJUNCTION EFFECT IN SILICON INTEGRATED CIRCUITS AND SENS0RS
Fruett and Meijer
ISBN: 1-4020-7053-5

CMOS CURRENT AMPLIFIERS; SPEED VERSUS NONLINEARITY
Koli and Halonen
ISBN: 1-4020-7045-4

MULTI-STANDARD CMOS WIRELESS RECEIVERS
Li and Ismail
ISBN: 1-4020-7032-2

A DESIGN AND SYNTHESIS ENVIRONMENT FOR ANALOG INTEGRATED CIRCUITS
Van der Plas, Gielen and Sansen
ISBN: 0-7923-7697-8

RF CMOS POWER AMPLIFIERS: THEORY, DESIGN AND IMPLEMENTATION
Hella and Ismail
ISBN: 0-7923-7628-5

DATA CONVERTERS FOR WIRELESS STANDARDS
C. Shi and M. Ismail
ISBN: 0-7923-7623-4

DIRECT CONVERSION RECEIVERS IN WIDE-BAND SYSTEMS
A. Parssinen
ISBN: 0-7923-7607-2

AUTOMATIC CALIBRATION OF MODULATED FREQUENCY SYNTHESIZERS
D. McMahill
ISBN: 0-7923-7589-0

MODEL ENGINEERING IN MIXED-SIGNAL CIRCUIT DESIGN
S. Huss
ISBN: 0-7923-7598-X

ANALOG DESIGN FOR CMOS VLSI SYSTEMS
F. Maloberti
ISBN: 0-7923-7550-5

CONTINUOUS-TIME SIGMA-DELTA MODULATION FOR A/D CONVERSION IN RADIO RECEIVERS L. Breems, J.H. Huijsing
ISBN: 0-7923-7492-4

CIRCUIT DESIGN FOR WIRELESS COMMUNICATIONS

Improved Techniques for Image Rejection in Wideband Quadrature Receivers

by

Kong-Pang Pun

The Chinese University of Hong Kong, Hong Kong

José Epifânio da Franca

ChipIdea Microelectronics S.A., Portugal

and

Carlos Azeredo-Leme

Instituto Superior Técnico, Lisbon, Portugal

SPRINGER-SCIENCE+BUSINESS MEDIA, B.V.

A C.I.P. Catalogue record for this book is available from the Library of Congress.

DOI 10.1007/978-1-4757-3737-0

Printed on acid-free paper

Preface

Eight decades ago, Armstrong invented the hyterodyne radio receiver architecture which was a great success that almost all high performance wireless receivers reported in the literature have adopted this architecture. Today, cellular communication systems are required to provide voice, data, video and audio communication services. Two key requirements for the design of a mobile terminal in such a system are low cost and low power consumption. The cost is widely concerned as the most important factor in today's very competitive environment. On the other hand, low power consumption is mandatory to prevent the shortening of battery life time while the amount of information to be processed by the terminal is steadily increasing. Many people believe that by integrating as many circuit components as possible in a CMOS technology can help to reduce the cost and power consumption, and at the same time, reduce the size of the receiver.

Hyterodyne architectures are not suitable for high level of integration, because they need many off-chip image rejection filters and channel selection filters. Recently, receiver architectures that are more suitable for high level of integration, such as the image rejection, low IF and direct conversion, have attracted much attention from many design engineers and researchers. All these architectures that apply image cancellation methods rather than off-chip image rejection filters encounter a fundamental problem of limited image rejection performance caused by analogue circuit imperfection such as the gain and phase imbalance between the in-phase (I) and quadrature (Q) paths of the receiver. This problem becomes more prominent if the receiver is of wideband. This limited image rejection is a big obstacle to achieve single chip integration of the receiver.

This book focuses on the image rejection problem and its solutions in various receiver architectures. Basically, the non-filtering methods for improving image rejection can be divided into two broad types. The first type of methods provides more accurate analog circuits that have less impact on the I and

Q imbalance. The second type of methods corrects or calibrates the I and Q imbalance by tuning or digital signal processing.

Apart from conventional methods, several new methods have been presented in this book. The new methods of the first type includes the switched-capacitor Hilbert transformers for accurate quadrature signal generation in a wide bandwidth and high performance sampling circuits that can also perform quadrature signal generation. The new methods of the second type includes the wideband digital I/Q imbalance calibration method and the adaptive I/Q imbalance correction method. These calibration or correction methods have taken the frequency dependence of the I/Q imbalance, which must be considered in a wideband receiver, into account. Some design examples have been included to demonstrate the proposed methods.

This book is mainly based on the materials of my PhD thesis which was supervised by Prof. Franca and co-supervised by Prof. Azeredo-Leme. The research work presented in this book was supported by the Foundation of Science and Technology of the Ministry of Science and Technology of Portugal and the European Commission ESPRIT project PAPRICA.

I would like to take this opportunity to thank my former colleagues in the Integrated Circuits and Systems Group, the Instituto Superior Téchnico and in Chipidea Microelectronics. They are Prof. João Vital, Prof. Nuno Horta, Eng. Ricardo Reis, Eng. Paulo Santos, Eng. Rui Neves, Eng. Nuno Garrido, Dr. João Goes, Dr. Kam-Wang Tam, Dr. Seng-Pan U, Eng. Nuno Franca, Dr. Ping Wang, Eng. Xiang Guan, Eng. Marco Oliveira, Dr. Yanyan Qiu Azeredo-Leme and Dr. Jingnan Xu. Their technical helps, discussions, experience sharing and friendship are most treasured by the author. Also, I am grateful to Dra. Ana Marcelino and Dra. Paula Silva for their helps from travelling arrangement to my visa renewal. I also thank Prof. Antonio Petraglia of Federal University of Rio de Janeiro for discussions on switched-capacitor Hilbert transformers. My gratitude also goes to Mr. Mark de Jongh of Kluwer Academic Publisher, and Prof. P.C. Ching, Prof. Oliver C.S. Choy and Prof. C.F. Chan of the Chinese University of Hong Kong, who have made the publication of this book possible.

Last but not least, I deeply thank my lovely wife who took over all my family responsibilities when I was studying abroad.

Kong-pang Pun,
Hong Kong, December 12, 2002.

Contents

List of Abbreviations

A/D	Analog-to-Digital
ADC	Analog-to-Digital Converter
AGC	Auto Gain Control
BB	Baseband
BPF	Band Pass Filter
CMOS	Complementary Metal Oxide Semiconductor
CT	Continuous-Time
CDS	Correlated Double Sampling
DC	Direct Current
DFT	Digital Fourier Transform
DIRS	Double Image Rejection Sampling
DQS	Double Quadrature Sampling
DSP	Digital Signal Processing
FIR	Finite Impulse Response
FLOPS	Floating Point Operations
GSM	Global System for Mobile communications
GBW	Gain-Bandwidth Product
I	In-phase
IC	Integrated Circuit
IF	Intermediate Frequency
IIR	Infinite Impulse Response
IIP3	Input-referred third order Intercept Point
IRR	Image Rejection Ratio
LMS	Least Mean Square
LNA	Low Noise Amplifier

LO	Local Oscillator
LPF	Low Pass Filter
MOSFET	Metal Oxide Semiconductor Field Effect Transistor
MOPS	Million Operation Per Second
NF	Noise Figure
NMOS	N-channel MOSFET
OPAMP	Operational Amplifier
OTA	Operational Transconductance Amplifier
PMOS	P-channel MOSFET
PSN	Phase Shifting Network
Q	Quadrature
Q-factor	Quality Factor
RF	Radio Frequency
SAW	Surface Acoustic Wave
SC	Switched Capacitor
SDIRRx	Sampled-Data Image Rejection Receiver
SFDR	Spurious Free Dynamic Range
S/H	Sample and Hold
SNR	Signal to Noise Ratio
VLSI	Very Large Scale of Integration

Chapter 1

Introduction

1.1 Motivations

In the last two decades, we have witnessed wireless communications evolving from the first generation analog systems to the second generation digital systems (Table 1.1), with dramatic down-scaling and price decreasing of the mobile terminals as well as longer stand-by time. This evolution has been enabled by significant advances in radio and integrated circuit techniques. For example, time-division or code-devision multiple access enabled by modern digital signal processing, together with the vary large scale integrated circuit (VLSI) increased significantly radio capacity and brought the radio costs down to the consumer level [1]. Today, we are seeing the emergence of the third generation wireless communication systems capable of transmitting various services from voice to multimedia (including voice, video, data, Internet, etc) [2, 3] with ever increased bandwidth and data rates.

Another point drawn our attention is that there is a strong need for multi-standard mobile terminals. As seen from Table 1.1, different standards have different bandwidth. To accommodate different bandwidth signals, a multi-standard receiver must have a bandwidth equal to the largest one. In another words, the receiver has to be wideband.

While the functionality of a mobile terminal is steadily increasing, there is a great challenge to prevent the shortening of battery life because the battery technology is not keeping pace with it. This challenge makes low power a key requirement in the mobile transceiver circuit design. The cost is obviously another key requirement [4].

High level of integration is widely considered as a way to achieve low-power and low-cost transceiver design [5, 6]. By increasing the integration level of a

Table 1.1: Characteristics of some mobile communication systems.

Standards	Multiple Access	Frequency Range	Channel Spacing	Modulation Scheme
AMPS	FDMA	869-894 MHz downlink† 824-849 MHZ uplink†	30 kHz	FM
PDC	TDMA	810-826 MHz downlink 940-956 MHZ uplink	25 kHz	$\pi/4$ DQPSK
GSM900	TDMA	935-960 MHz downlink 890-915 MHZ uplink	200 kHz	0.3 GMSK
DCS1800	TDMA	1805-1880 MHz downlink 1710-1785 MHZ uplink	200 kHz	0.3 GMSK
DECT	TDMA	1897-1913 MHz	1728 kHz	0.5 GFSK
IS-54	TDMA	869-894 MHz downlink 824-849 MHZ uplink	30 kHz	$\pi/4$ DQPSK
IS-95	CDMA	869-894 MHz downlink 824-849 MHZ uplink	1250 kHz	QPSK
UMTS	WCDMA	2110-2170 MHz downlink 1920-1980 MHz uplink	5 MHz	QPSK

† Downlink: from the base station to the mobile station; Uplink: the reversed direction.

mobile transceiver, the parasitics are dramatically reduced so the power dissipation will be lowered. The cost will be reduced because the component count is reduced and consequently, the assembling and testing procedures are reduced. Lastly, high level of integration can also reduce the size of the transceiver.

A fundamental challenge to the high level integration of a radio receiver in the architectural level is the *image problem*. Figure 1.1 illustrates the problem. Two radio frequency (RF) input signals at frequencies of $(\omega_{LO} + \omega_{IF})$ and $(\omega_{LO} - \omega_{IF})$ will be down-converted to the same intermediate frequency (IF) ω_{IF} by mixing them with a local oscillator at frequency of ω_{LO}. One of the inputs could be the desired signal and the other is referred as the *image interferer*. They are apart from each other by $2\omega_{IF}$, and look like an image of each other with respect to the local oscillator frequency ω_{LO}. This phenomenon can be blamed at the fact that the real RF and LO signal have spectral components at both positive and negative frequencies. The image interferer must be rejected to prevent aliasing with the desired signal, and this causes problems in achieving high level integration of the receiver.

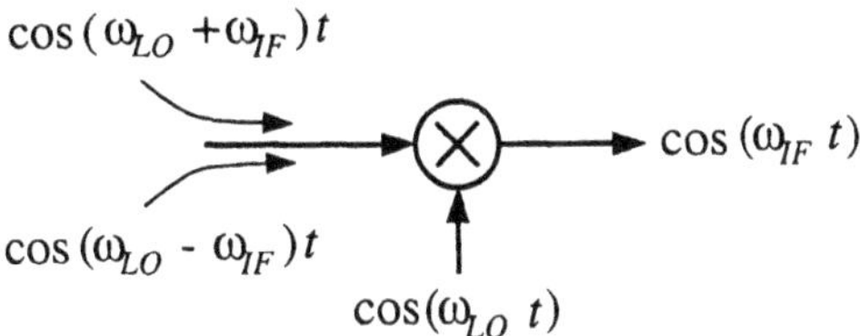

Figure 1.1: The image problem.

Image rejection methods can be divided into two classes: pre-mixer filtering method and complex mixing (or quadrature mixing, vector mixing) method.

The widely-used classical super-heterodyne architecture [7, 8, 9, 10] employs the pre-mixer filtering method. As a super-heterodyne receiver has several mixer stages, it requires several image rejection filters. This filters are accurate frequency selective filters that are very hard to integrate by today's technology. Usually, a super-heterodyne uses several off-chip filters for the purposes of image rejection as well as channel selection.

An eminent example of using complex mixing to reject the image is the direct conversion architecture [11, 12, 13] which can achieve very high level of receiver integration. The receiver employs in-phase (I) and quadrature (Q) mixers to perform complex down-conversion of RF signal to DC. In this case, the image frequency corresponds the negative of the desired signal frequency, and the image interferer is referred as the *self-image.*

By complex down-conversion, or quadrature demodulation equivalently, the receiver is able to reject the image in principle. However, practical circuit imperfections like mismatches between the I and Q channels will limit the receiver's image rejection performance. This limitation is usually within the tolerable range since the image magnitude equals the desired signal magnitude. However, things get worse when one moves the channel selection function from analog to digital domain in a digital receiver. This means that several radio channels co-exist after the analog-to-digital conversion. So a strong signal can be the image of a week signal and the image rejection requirement becomes very high which requires highly-matched I and Q channels.

Besides, the direct conversion architecture has drawbacks of LO leakage and the DC offset and low frequency noise that are just inside the baseband [11]. To avoid these drawbacks while maintaining the advantages of direct conversion architecture, the intermediate frequency of the receiver can be placed at a low but non-zero frequency [14, 15]. Again, the RF signal must be down-converted by complex mixer to preserve the image rejection ability. This type of receivers can perform analog-to-digital conversion at IF stage. The IF-to-baseband conversion and channel selection can be performed by digital circuitry [16, 17, 18, 19].

Different from the single channel direct conversion receiver, the low-IF receivers can have an image interferer much stronger than the desired signal. For example, blocking interferers as high as 80 dBc and just 3 MHz away from the desired signal must be rejected in a GSM receiver. The image rejection ratio of a receiver with a typical 1% channel mismatch is about 40 dB, which is far from the requirement.

The low-IF architecture is actually a variation of the traditional image rejection receiver [20] whose output is real instead of complex. Again, due to the gain mismatch and phase errors between the internal I and Q paths, the image rejection performance of the traditional image rejection receiver is limited.

There are methods to improve the image rejection performance of the receivers mentioned above, for example, digital or analog calibration of I/Q mismatches [21, 22], the use of digital compensation of I/Q mismatches [23], the use of on-chip passive or active image rejection filters [24], and the use of circuit components with better accuracy [25], etc. However, the reported methods are mainly applicable for narrow-band systems (where the I/Q mismatches are assumed to be frequency-independent) only, or have limited performance that an external RF image rejection filter is still required.

Our primary motivation is to find innovative and effective methods to conquer the image problem so that to contribute to the higher level integration of wideband radio receivers for for today's and tomorrow's radio communication systems.

1.2 Objectives and Approaches

The general objectives of this book are to study the image problems associated with wideband quadrature receivers of various architectures, and to provide effective solutions without using off-chip filters so that higher receiver integration can be achieved.

As the image problem is closely related to the gain and phase imbalances between the I and Q paths of the receiver, three approaches of the following will be taken to achieve our objectives:

- To find receiver circuits with better I/Q matching performance;
- To find circuit architectures that are less sensitive to I/Q imbalance;
- To correct the I/Q imbalances by digital signal processing methods.

Particular efforts will be devoted to the receivers of image-reject, low-IF and direct conversion architectures.

1.3 Book Outline

This book is organised in eight chapters. Each chapter is arranged to be self-contained as much as possible. The remaining chapters are outlined below.

Chapter two gives a general overview on various radio receiver architectures, including the super-heterodyne, zero-IF, low-IF, image-reject receivers and software radio, with the focus on the image problems associated with each architecture. Existed image rejection solutions and their limitations are briefed.

Chapter three focuses on an important functional block, the 90° phase shifter, which has a significant impact on the image rejection performance of a quadrature radio receiver. Traditional realisations, including the passive and active RC-CR circuits, polyphase networks, etc., are presented first. Then the discrete-time implementations, i.e, the switched-capacitor (SC) Hilbert transformers are discussed. The design of FIR and IIR Hilbert transformers is also mentioned. High performance SC Hilbert transformers with low sensitivity to finite amplifier gain and bandwidth are introduced.

Chapter four addresses a novel sampled-data image-rejection receiver architecture, which employs an SC IIR Hilbert transformer as the accurate wideband 90° phase shifter for better image rejection. A prototype chip targeted to the application of cordless telephones is realized in $0.6\mu m$ CMOS technology. Experimental results are reported.

Chapter five deals with the image rejection problem in low-IF receivers with direct IF digitising. Lowpass delta sigma modulators with integrated mixer for the IF A/D conversion are discussed in details. The effect of channel mismatches and phase errors of the integrated mixer is analysed. Various methods for improving the image rejection in these IF A/D converters are introduced, including a mismatch and phase error free IF-to-baseband mixing circuit.

Chapter six and seven present digital methods to correct the I/Q imbalances in direct conversion and low IF receivers by calibration and blind signal processing approaches, respectively. The issues in wideband and multi-channel reception are discussed.

Chapter six includes Churchill's and statistical methods for frequency independent I/Q mismatch calibration, and a wideband method capable of calibrating frequency dependent I/Q mismatches. The blind signal processing approach presented in Chapter seven is also capable of correcting frequency dependent I/Q mismatches, but needs no external reference signal.

Chapter eight concludes the overall research work presented in this book. Perspectives for future work are pointed out.

References

[1] C.K. Coursey, *Understanding digital PCS, the TDMA Standard*, Artech House, 1999.

[2] Malcolm W. Oliphant, "The mobile phone meets the internet," *IEEE Communications Magazine*, pp. 20–28, Aug. 1999.

[3] William Sweet, "Cell phones answer internet's call," *IEEE Spectrum*, pp. 42–46, Aug. 2000.

[4] P. Gray and R. Meyer, "Future directions of silicon ICs for RF personal communications," in *Custom Integrated Circuits Conference*, 1995, pp. 83–90.

[5] J.C. Rudell, J.J. Ou, et al., "Recent developments in high integration multi-standard cmos transceivers for personal communication systems," *Int. Sym. on Low Power Electronics, Monterey, California*, 1998.

[6] A. Abidi et al., "The future of CMOS wireless transcivers," in *Digest of Technical Papers, IEEE Int. Solid-State Circuit Conference*, Feb. 1997, pp. 118–119.

[7] V. Thomas et al., "A one-chip 2 GHz single-superhet receiver for 2Mb/s FSK radio communications," in *Digest of Technical Papers, IEEE Int. Solid-State Circuit Conference*, San Francisco, CA, Feb. 1994, pp. 42–43.

[8] T.D. Stetzler, I.G. Post, J.H. Havens, and M. Koyama, "A 2.7-4.5V single chip GSM transceiver RF integrated circuit," *IEEE J. Solid-State Circuits*, vol. 30, no. 12, pp. 1421–1429, Dec 1995.

[9] K. Irie, H. Matsui, T. Endo, et al., "A 2.7V GSM RF transceiver IC," in *Digest of Technical Papers, IEEE Int. Solid-State Circuit Conference*, Feb. 1997, pp. 302–303.

[10] P. Orsatti, F. Piazza, Q. Huang, and T. Morimoto, "A 20mA-receive 55mA-transmit GSM transceiver in 0.25μm CMOS," in *Digest of Technical Papers, IEEE Int. Solid-State Circuit Conference*, 1999, pp. 232–234.

[11] A. A. Abidi, "Direct-conversion radio transceivers for digital communications," *IEEE J. Solid-State Circuits*, vol. 30, no. 12, pp. 1399–1410, Dec. 1995.

[12] T. Tsukahara, M. Ishikawa, and M. Muraguchi, "A 2V 2GHz Si-bipolar direct-conversion quadrature modulator," *IEEE J. Solid-State Circuits*, vol. 31, no. 2, pp. 262–267, Feb 1996.

[13] J. Tang and D. Kasperkovitz, "A 0.9-2.2GHz monolithic quadrature mixer oscillator for direct-conversion satellite receivers," in *Digest of Technical Papers, IEEE Int. Solid-State Circuit Conference*, Feb. 1997, pp. 88–89.

[14] J. Crols and M. Steyaert, "A 1.5GHz highly linear CMOS down conversion mixer," *IEEE J. Solid-State Circuits*, vol. 30, no. 7, pp. 736–742, July 1995.

[15] J.C. Rudell, J.J. Ou, et al., "A 1.9GHz wide-band IF double conversion CMOS receiver for cordless telephone application," *IEEE J. Solid-State Circuits*, vol. 32, pp. 2071–2088, Dec 1997.

[16] H.J. Dressler, "Interpolative bandpass A/D conversion - experimental results," *IEE Electron. Letters*, vol. 26, no. 20, pp. 1652–1653, Sept. 1990.

[17] A.M. Thurston, T.H. Pearce, and M.J. Hawksford, "Bandpass implementation of the sigma-delta A-D conversion technique," *Proc. IEE Int. Conference on A/D and D/A Conversion, Swansea, U.K.*, pp. 81–86, Sept. 1991.

[18] S.A. Jantzi, W.M. Snelgrove, and P.F. Ferguson Jr., "A fourth-order bandpass sigma-delta modulator," *IEEE J. Solid-State Circuits*, vol. 28, no. 3, pp. 282–291, March 1993.

[19] S. Jantzi, R. Schreier, and M. Snelgrove, "The design of bandpass $\Delta\Sigma$ ADCs," in *Delta-Sigma Data Converters, Theory, Design and Simulation*, S. Norsworthy, R. Schreier, and G.C. Temes, Eds., pp. 282–308. IEEE Press, 1997.

[20] D.K. Weaver, "A third method of generation and detection of single-sideband signals," *Proc. IRE*, vol. 44, pp. 1703–1705, Dec 1956.

[21] Behazad Razavi, "Design consideration for direct-conversion receivers," *IEEE Trans. on Circuits and Systems - II: Analog and Digital Signal Processing*, vol. 44, no. 6, pp. 428–435, June 1997.

[22] F.E. Churchill, G.W. Ogar, and B.J. Thompson, "The correction of I and Q errors in a coherent processor," *IEEE Transactions on Aerospace and Electronic Systems*, vol. AES-17, no. 1, pp. 131–137, Jan 1981.

[23] Li Yu and W. M. Snelgrove, "A novel adaptive mismatch cancellation system for quadrature IF radio receivers," *IEEE Transactions on Circuits and Systems: - II: Analog and digital signal processing*, vol. 46, no. 6, pp. 789–801, June 1999.

[24] J. Crols and M. Steyaert, "An analog integrated polyphase filter for a high performance low-if receivers," in *Proc. VLSI Circuits Symposium*, Kyoto, June 1995, pp. 87–88.

[25] T. Okanobu, H. Tomiyama, and H. Arimoto, "Advanced low voltage single chip radio IC," *IEEE Trans. Consumer Electronics*, vol. 38, no. 3, pp. 465–475, August 1992.

[26] E. van der Zwan, K. Philips, and C. Bastiaansen, "A 10.7MHz IF-to-basebad $\Delta\Sigma$ A/D conversion system for AM/FM radio receivers," in *Digest of Technical Papers, IEEE Int. Solid-State Circuit Conference*, Feb. 2000, pp. 340–341.

Chapter 2

Wireless Receivers: Architectures and Image Rejection

2.1 Introduction

As early as Armstrong invented the *heterodyne* receiver architecture eight decades ago [1], the image rejection had emerged as an important issue in the design of a radio receiver. The image problem arises from the fact that radio interferer at the *image frequency* will be downconverted to the same intermediate frequency (IF) as the desired signal and therefore corrupt it. The traditional method for rejecting the image interferer is to use a high quality factor (Q-factor) band-pass filter before the RF mixer. At that time all the electrical components were discrete, so was the image-reject filter. Currently, a majority of those discrete components can be put together to a small integrated circuit die, but hardly the image-reject filters. For high level receiver integration, this approach is not favoured.

The second approach to reject the image is to employ a complex mixer, or image-reject mixer, which does not need a filter in principle. Receivers of this type include Harley and Weaver receivers. However, due to practical analogue circuit imperfections, mainly the I and Q channel imbalances, these receivers provide insufficient image rejection performance. We will discuss this issue in details.

Almost all modern radio receivers employ quadrature modulation/demodulation schemes. We start this chapter by explaining why the quadrature modulation schemes are used, especially in digital communication systems. Then we present various receiver architectures including the heterodyne, homodyne, image-reject and low-IF receivers with focus on their image rejection problem. Last, the new concept of software radio which entails no image problem is briefly discussed.

2.2 Why use Quadrature Modulation?

A physical radio signal is always a real signal. But a modern receiver normally demodulates the signal to an in-phase (I) component and a quadrature (Q) component as shown in Figure 2.1. The quadrature demodulation is performed by an I/Q mixer which uses two local oscillators with a same frequency but a 90° phase difference. The I and Q components of the demodulated signal are independent and orthogonal to each other. They carry different information. One can be changed without affecting the other.

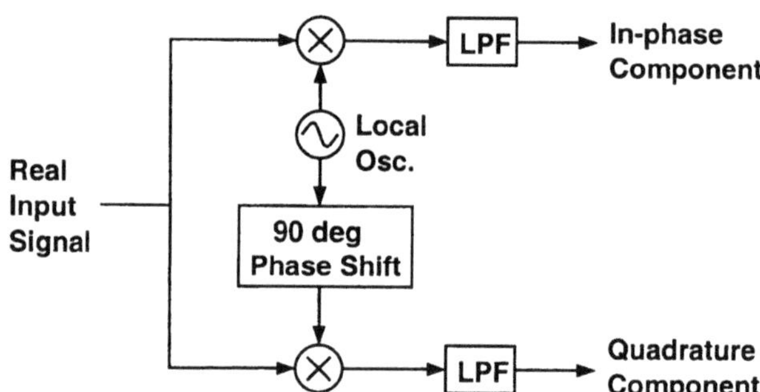

Figure 2.1: I and Q demodulator.

Obviously more hardware is needed to carry out the quadrature modulation /demodulation. But it is still desirable for the following reasons:

- First, the bandwidth of the input signal can be doubled if both outputs are digitised. This point can be explained in either the time or the frequency domain. In the time domain, if the sampling frequency is f_s, one must obtain two samples per cycle at the highest input frequency to fulfil the Nyquist sampling theory; thus, the highest frequency is $f_s/2$. If there is a Q channel, two more samples will be collected; thus, the highest frequency can be extended to f_s. In the frequency domain, if the input is real there are positive and negative frequency components, and the highest frequency without ambiguity is $f_s/2$. For complex data, there are only positive or negative frequencies, and the unambiguous range extends to f_s.

- Second, digital modulation is easy to accomplish with I/Q modulators. Most digital modulation maps the data to a number of discrete points on the I/Q plane. These are known as constellation points [2]. As the signal moves from one point to another, simultaneous amplitude and phase modulation usually results. To accomplish this with an amplitude modulator and a phase modulator is difficult and complex. It is also impossible with a conventional phase modulator. The signal may, in principal, circle the origin in one direction forever, necessitating infinite phase shifting capability. Alternatively, simultaneous amplitude and phase modulation is easy with an I/Q modulator. The I and Q control signals are bounded, but infinite phase wrap is possible by properly phasing the I and Q signals.

2.3 Heterodyne Receiver

Heterodyne receiver was invented by Armstrong in 1918 [1]. It is generally thought to be the receiver of choice. Something like 98% of radio receivers use this architecture [3]. Examples can be found in [4, 5, 6, 7].

Figure 2.2 shows a single IF heterodyne receiver which can be divided to two stages. The first stage consists of a duplexer, a low noise amplifier (LNA), an RF image-reject filter (RF bandpass filter) and an RF mixer with a local oscillator. The second stage consists of a channel-select filter (IF bandpass filter), an auto gain control (AGC) unit and an I and Q demodulator as shown in Figure 2.1. Among these, RF and IF bandpass filters are usually off-chip, like surface acoustic wave (SAW) or ceramic filters [8, 9, 10].

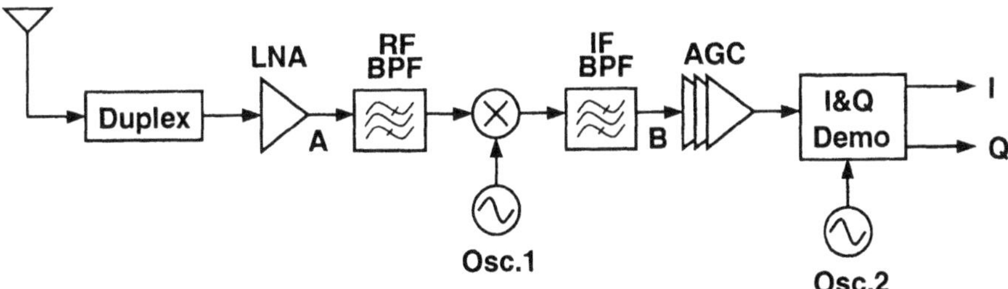

Figure 2.2: A typical heterodyne receiver architecture.

Figure 2.3 displays the frequency spectra at different points of the heterodyne receivers. First, the input signal is shifted to a lower frequency f_{IF} by the RF mixer, then passes through a channel-select filter to remove the adjacent interferers. Finally, the channel-selected signal is demodulated into I and Q components in baseband. Note that the output spectrum is asymmetric due to its complex nature.

Since its center frequency is low, the channel-select filter has much more relaxed requirements than if it is implemented at RF stage. For example, for an intermediate frequency of 10 MHz and a channel spacing of 200 kHz, a Q-factor of only 50 is required for the filter. Therefore, very good selectivity can be obtained easily. Another advantage of low IF is that the I and Q mismatch in the quadrature demodulator is more easily controlled.

The lower is the intermediate frequency, the more relaxed is the requirement of the channel-select filter. But this increases the difficulty to reject the image interferer. To understand this problem, suppose the RF signal is $\cos(\omega_{RF}t)$, and the LO signal is $\cos(\omega_{LO}t)$, where $\omega_{RF} - \omega_{LO} = \omega_{IF}$. Multiplying these two signals results:

$$\cos(\omega_{RF}t)\cos(\omega_{LO}t) = (1/2)\left[\cos(\omega_{IF}t) + \cos(\omega_{RF} + \omega_{LO}t)\right]. \qquad (2.1)$$

The wanted signal is downconverted to IF. Multiplying the LO with an interferer $\cos(\omega_{img}t)$, where $\omega_{img} = w_{LO} - w_{IF}$ is referred as the image frequency, we obtain:

$$\begin{aligned} & \cos(\omega_{img}t)\cos(\omega_{LO}t) \\ = \; & (1/2)\left[\cos(\omega_{img} - \omega_{LO})t + \cos(\omega_{LO} + \omega_{img})t\right] \\ = \; & (1/2)\left[\cos(\omega_{IF}t) + \cos(2\omega_{LO} - \omega_{IF})t\right]. \end{aligned} \qquad (2.2)$$

An output component with the same IF as the desired signal is resulted. This effect is referred as *image aliasing.*

In certain special circumstance, the aliased image interferer can be separated from the desired signal though a careful selection of the intermediate frequency. A typical example is the standard broadcast FM receiver. In such a receiver, the 10.7 MHz IF guarantees that the image channel lies outside the 20 MHz wide FM band. Therefore, the subsequent frequency-discriminating detector will inherently tend to reject the image signal which is assumed not an FM signal.

However, in general the image signal can not be distinguished from the desired signal, and must be removed before it is downconverted. Traditionally, this is done by an off-chip RF *image-reject filter.* The image-reject filter has a center frequency of f_{RF}, and must suppress the image at a distance of $2f_{IF}$. It is

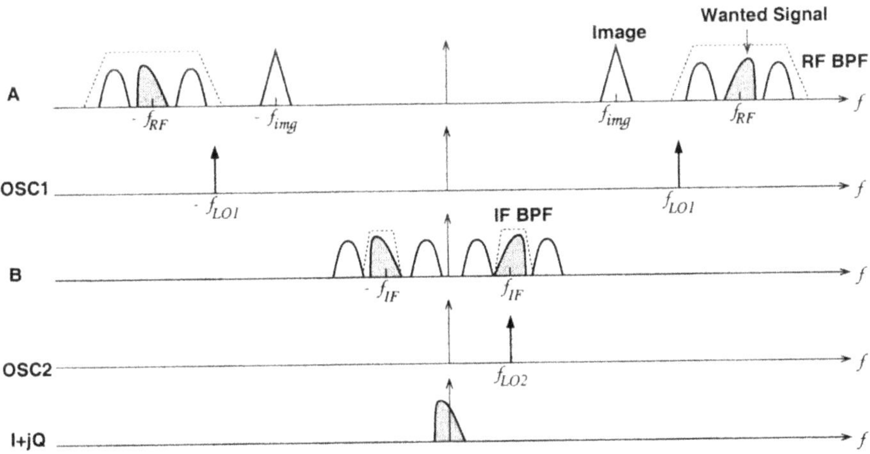

Figure 2.3: The frequency spectra at different points of the heterodyne receiver.

obvious that decreasing the intermediate frequency will tighten the requirement of the image reject filter. So there is a trade-off in choosing the intermediate frequency. In the case that the input frequency in very high, for example 900 MHz in GSM, more IF stages are usually adopted to solve this dilemma [11, 12, 13, 14] with the cost of more hardware, including more off-chip image-reject filters.

Monolithic integration of image-reject filter draws more and more attention recently. By monolithic integration the expensive SAW filter can be eliminated. Moreover, the LNA does not need to drive 50 Ω load and the mixer does not need to exhibit 50 Ω input impedance anymore. This can release a lot of room for optimising power consumption, noise figure (NF), gain and other important design parameters of LNA and mixer. While very high-Q BPF is almost impossible to integrate, a notch filter is possible. In [15, 16], on-chip LC tank was exploited to place a notch at the image frequency. In [17], an inductor-less CMOS notch filter was suggested. Performance parameters of these notch filters are listed in Table 2.1 and are compared with those of a commercial RF SAW filter for a GSM mobile receiver. From the table, it can be found that the notch filters have good performance in image rejection, but have disadvantages in NF, linearity and power consumption (in inductor-less filter). Another drawback of these notch filters is that frequency tuning is required.

Table 2.1: Performance of different image-reject filters.

	SAW filter [18]	Integrated LC notch filter [16]	Integrated active notch filter [17]†
Passband	935-960 MHz	1900 MHz	947 MHz
NF/Insertion loss	3.1 dB	4.8 dB	7.2 dB
Image Rejection	50 dB @+140 MHz	65 dB @+600 MHz	60dB @+140 MHz
IIP3	-	-19 dBm	-20 dBm
Power consumption	-	-	27 mW

† Simulation results.

The advantages of heterodyne receiver are summarised as follows: (1) selectivity is very good; (2) requirements on the channel selection filter is low; (3) DC offset of the first few stages is eliminated by the BPF; and (4) I-Q mismatch occurs at low frequency and is easier to control and correct. The main drawback is that high-Q image reject filters are required. This makes it very hard to achieve full integration.

2.4 Image-Reject Receivers

As mentioned in the beginning of this chapter, the second approach to reject the image is to employ a complex mixer, or *image-reject mixer* as traditionally called. A receiver employing an image-reject mixer is called *image-reject receiver.*

The primary advantage of the image rejection receivers is that they do not need image-reject filters. Without the image-reject filters, the intermediate frequency can be placed very low as the trade-off between the requirement on the image rejection filter and the channel-select filter discussed in the previous section no longer exits. Therefore, good selectively could be achieved by integrated filters with a Q-factor of only 10 to 20, typically.

The operational principles and practical limitation of the two traditional image-reject receivers, namely, Hartley and Weaver receivers, are discussed below.

2.4.1 Hartley architecture

This architecture was proposed by Hartley [19] in 1928. Figure 2.4 shows a block diagram of this receiver. It consists of two matched mixers, a 90° phase shifter, a pair of LO with 90° phase difference, and an adder. An I/Q demodulator as shown in Figure 2.1 can be added at the end to produce quadrature baseband outputs.

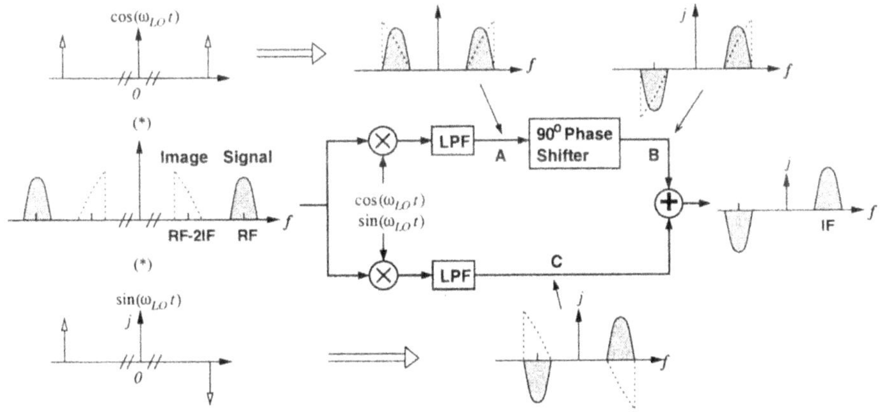

Figure 2.4: Principle of Hartley image-reject receiver.

Figure 2.4 shows also the frequency spectra at different points of the Hartley receiver. Here, the quadrature signal is defined to leg the in-phase signal by 90°, and phase shifting is defined to shift phase in advance. In frequency domain, 90° phase shifting corresponds to multiplying positive and negative frequency spectrum by j and $-j$ respectively. And the operation of mixing corresponds to convolving the input spectrum with the LO spectrum.

In Figure 2.4, $\omega_{LO} < \omega_{RF}$ is assumed. Therefore the image frequency is equal to $2\omega_{LO} - \omega_{RF}$. In this receiver, the desired signal and the image interferer are downconverted together in both upper and lower paths. However, the desired signals at the end of the upper and the lower paths are in-phase, while the image interferers are 180° out of phase. When the upper and the lower paths are recombined, the image interferer will be cancelled out and the desired signal will be left.

The image cancellation can be also explained in the time domain. Suppose that the wanted signal and image interferer, I and Q phases of LO signal are $A\cos(\omega_{RF}t)$, $B\cos(\omega_{img}t)$, $\cos(\omega_{LO}t)$ and $\sin(\omega_{LO}t)$, respectively, where $\omega_{RF} - \omega_{LO} = \omega_{LO} - \omega_{img}$. For a perfectly matched upper and lower paths, the signal at node A is

$$\begin{aligned}
&A\cos(\omega_{RF}t)\cos(\omega_{LO}t) + B\cos(\omega_{img}t)\cos(\omega_{LO}t)\\
&= A/2\cos(\omega_{RF} - \omega_{LO})t + B/2\cos(\omega_{img} - \omega_{LO})t \quad \text{(lowpassed)}\\
&= A/2\cos(\omega_{IF}t) + B/2\cos(\omega_{IF}t),
\end{aligned} \tag{2.3}$$

at node B is

$$\begin{aligned}
&A/2\cos(\omega_{IF} + 90^o) + B/2\cos(\omega_{IF}t + 90^o)\\
&= -A/2\sin(\omega_{IF}t) - B/2\sin(\omega_{IF}t),
\end{aligned} \tag{2.4}$$

and at node C is

$$\begin{aligned}
&A\cos(\omega_{RF}t)\sin(\omega_{LO}t) + B\cos(\omega_{img}t)\sin(\omega_{LO}t)\\
&= -A/2\sin(\omega_{IF}t) + B/2\sin(\omega_{IF}t) \quad \text{(lowpassed)}.
\end{aligned} \tag{2.5}$$

Summing up signals at node B and C, we obtain the output as $-A\sin(\omega_{IF}t)$ while the image term with coefficient B is eliminated.

Now consider the presence of gain mismatch and phase imbalance. To simplify the analysis, let us assign all these errors to the I and Q phases of LO signal. If these errors in each block are independent of frequency, then they have the same effects and this assignment does not lose the generality. Suppose the I and Q phases of LO signal are $(1+\alpha)\cos(\omega_{LO}t)$ and $\sin(\omega_{LO}t+\varepsilon)$ respectively, where α and ε are gain and phase errors respectively. The signal at node A becomes:

$$\begin{aligned}
&A\cos(\omega_{RF}t)(1+\alpha)\cos(\omega_{LO}t) + B\cos(\omega_{img}t)(1+\alpha)\cos(\omega_{LO}t)\\
&= (1+\alpha)\left[A\cos(\omega_{RF} - \omega_{LO})t + B\cos(\omega_{img} - \omega_{LO})t\right]/2\\
&= (1+\alpha)\left[A\cos(\omega_{IF}t) + B\cos(\omega_{IF}t)\right]/2 \quad \text{(lowpassed)}.
\end{aligned} \tag{2.6}$$

The signal at node B becomes:

$$\begin{aligned}
&(1+\alpha)\left[A\cos(\omega_{IF} + 90^o) + B\cos(\omega_{IF}t + 90^o)\right]/2\\
&= (1+\alpha)\left[-A\sin(\omega_{IF}t) - B\sin(\omega_{IF}t)\right]/2.
\end{aligned} \tag{2.7}$$

The signal at node C becomes:

$$\begin{aligned}
&A\cos(\omega_{RF}t)\sin(\omega_{LO}t+\varepsilon) + B\cos(\omega_{img}t)\sin(\omega_{LO}t+\varepsilon)\\
&= -A/2\sin(\omega_{IF}t - \varepsilon) + B/2\sin(\omega_{IF}t + \varepsilon) \quad \text{(lowpassed)}\\
&= -A/2[\sin(\omega_{IF}t)\cos\varepsilon - \cos(\omega_{IF}t)\sin\varepsilon]\\
&\quad +B/2[\sin(\omega_{IF}t)\cos\varepsilon + \cos(\omega_{IF}t)\sin\varepsilon]
\end{aligned} \tag{2.8}$$

Summing up the signal at B and C, we obtain the output as:

$$A/2[-(1+\alpha+\cos\varepsilon)\sin(\omega_{IF}t)-\cos(\omega_{IF}t)\sin\varepsilon]+ \\ B/2[(\cos\varepsilon-1-\alpha)\sin(\omega_{IF}t)+\cos(\omega_{IF}t)\sin\varepsilon], \tag{2.9}$$

where the first term is the desired signal and the second term is the image. Therefore, a residual image exists. From (2.9), we have the power of the desired signal as:

$$A^2/8\left[(1+\alpha)^2+2(1+\alpha)\cos\varepsilon+1\right], \tag{2.10}$$

and the power of residual image as:

$$B^2/8\left[(1+\alpha)^2-2(1+\alpha)\cos\varepsilon+1\right]. \tag{2.11}$$

Note that in the above equations A and B are used only for distinguishing the desired signal and the image. Normalise A and B to unity and divide (2.11) by (2.10), we obtain the *image rejection ratio* (IRR) of the receiver as:

$$IRR=\frac{1+2e_g\cos\varepsilon+e_g^2}{1-2e_g\cos\varepsilon+e_g^2}, \tag{2.12}$$

where $e_g = 1+\alpha$ is the gain ratio of the two paths. Figure 2.5 is a contour plot of the IRR as a function of gain and phase error. To improve the IRR, one may just need to improve the gain or phase error whichever is dominant.

Factors which limit the IRR in a integrated receiver includes the gain difference between two mixers, the magnitude imbalance and phase error between the quadrature outputs of LO, and the magnitude imbalance and phase error of the phase shifter in the signal path.

In the Hartley receiver, the source of mixer gain mismatch arises from local variations and alignment errors in the fabrication. Careful layout is necessary. To achieve higher gain matching, external tunable mixer could be employed [20]. However, this method is not preferred because the tuning procedure increases the production cost.

Quadrature output of LO can be generated by three methods: (1) use of oscillators with inherent quadrature outputs; (2) use of frequency divider with one output triggered by the rising edge and another by the falling edge; (3) use of phase shifter. The first method is most favourable. A phase error of 0.5^o and magnitude imbalance of 1% [21](corresponding to 45 dB IRR) can be easily achieved by this method. The performance of the second method is limited by the duty cycle of the clock signal [20]. Besides, this method is not suitable for high frequency application, because it needs a clock signal with doubled frequency. The third method is suitable only for systems with a narrow tuning range of the LO because the phase shifter has limited bandwidth [22, 5].

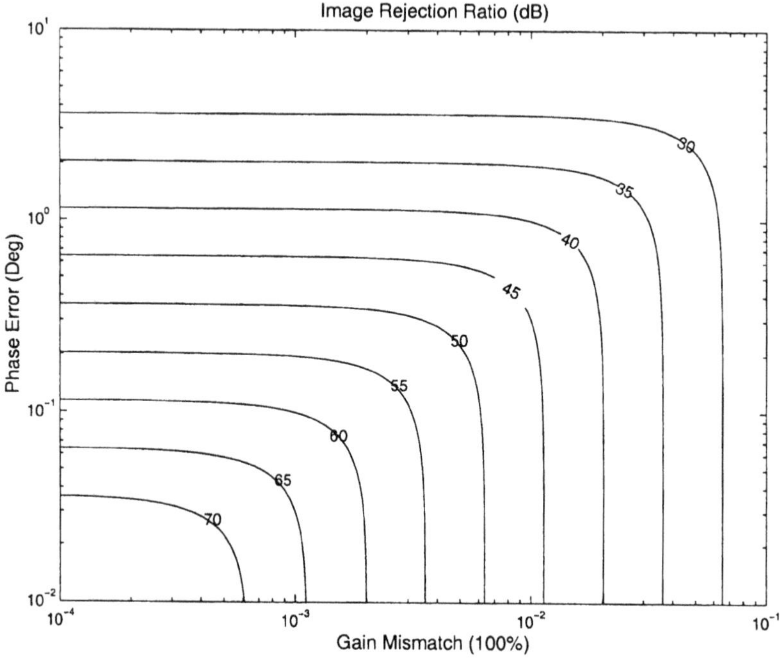

Figure 2.5: IRR versus gain and phase error.

A more critical problem is the phase error and magnitude imbalance generated by the 90° phase shifter in the signal path. All existing implementations of this phase shifter are based on the passive or active RC/CR circuits. The problem is that R and C is varying with temperature or process. It is difficult to achieve high magnitude balance in a broad frequency band. More details regarding the phase shifter will be discussed in Chapter three.

Table 2.2 list some reported IRR performance of Hartley receivers. The best IRR listed is 35 dB. Sometimes, an RF image-reject filter with relaxed Q-factor is still needed to help improve the IRR. In spite of the insufficient IRR, the Hartley architecture still finds many applications in wireless systems [20, 23, 24].

Figure 2.6 shows variations of Hartley architecture. Figure 2.6(a) and (b) are for input frequency higher than LO frequency, (c) and (d) are for input frequency lower than LO frequency. It is also possible to place the 90° phase shifter in the RF stage, i.e., before the mixer, instead of in the IF stage. This

Table 2.2: Reported IRR of integrated image rejection receivers

Author	IRR(dB)	Phase Shifter
McDonald [25]	14.1	Passive RC/CR
Baum [26]	34	Passive RC/CR
Pache [27]	35	Passive RC/CR
Okanobu [20]	30	Active RC/CR

approach has been exploited in the *double quadrature downconverter* which will be discussed later in this chapter.

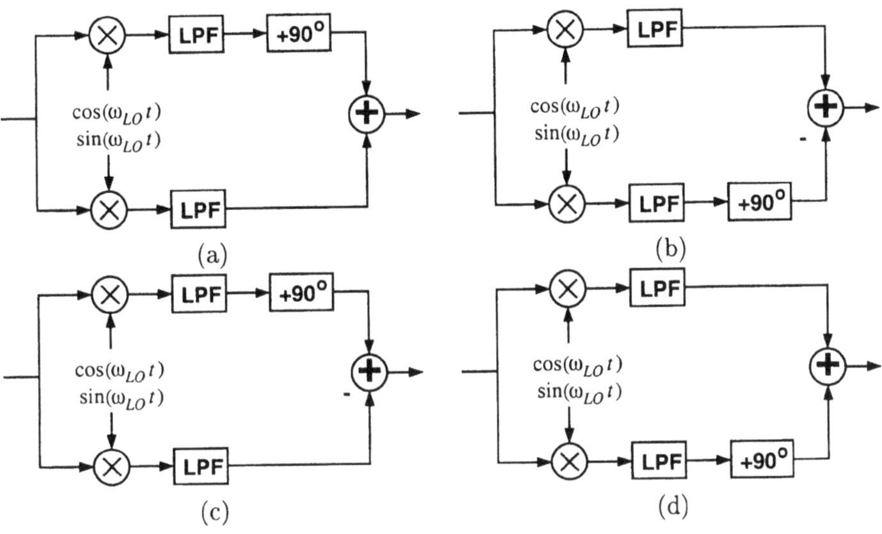

Figure 2.6: Variations of the Hartley receiver architecture: (a)-(b) for $\omega_{RF} > \omega_{LO}$; (c)-(d) for $\omega_{RF} < \omega_{LO}$;

2.4.2 Weaver architecture

The Weaver architecture [28] was invented in 1956. As shown in Figure 2.7, this architecture differs from the Hartley architecture in that the 90° phase shifting in the signal path is replaced by another pair of quadrature mixers. The purpose of this replacement is to perform phase shifting not on the wideband signal path, but on the second LO which is just a single sinusoidal tone. Therefore, phase shifting accuracy can be better controlled.

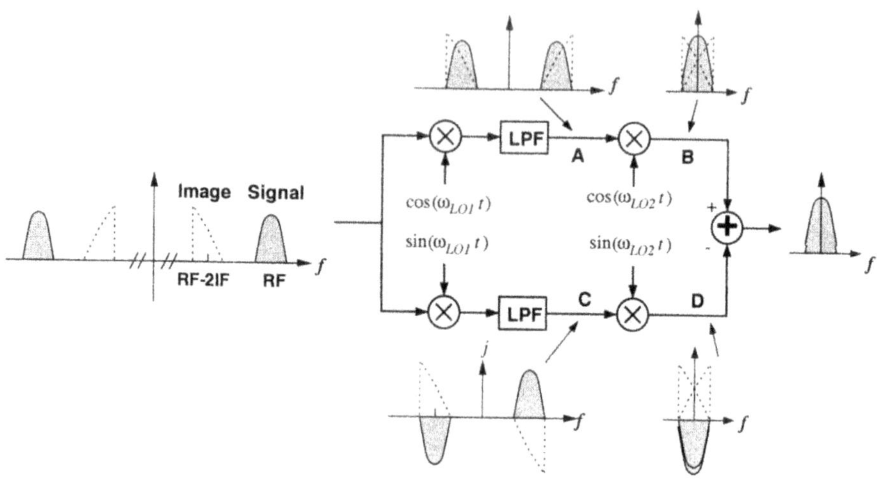

Figure 2.7: Principle of Weaver image-reject receiver.

Figure 2.7 shows also the frequency spectra at different nodes. The principle of image cancelling can be clearly understood from the frequency domain information shown in this figure. The configuration of Figure 2.7 has the output located at baseband.

Now, let us analyse the image cancelling process in time-domain. At notes A and C, the signals are same as those in Hartley receiver, i.e.,$A/2\cos(\omega_{IF}t) + B/2\cos(\omega_{IF}t)$ and $-A/2\sin(\omega_{IF}t) + B/2\sin(\omega_{IF}t)$, respectively (see (2.3)(2.5)). The signal at node B is:

$$\begin{aligned} & A/2\cos(\omega_{IF}t)\cos(\omega_{LO2}t) + B/2\cos(\omega_{IF})t\cos(\omega_{LO2}t) \\ = \; & A/4\cos(\omega_{IF} - \omega_{LO2})t + B/4\cos(\omega_{IF} - \omega_{LO2})t \quad \text{(lowpassed)} \end{aligned} \tag{2.13}$$

and the signal at node D is

$$\begin{aligned} & -A/2\sin(\omega_{IF}t)\sin(\omega_{LO2}t) + B/2\cos(\omega_{IF}t)\sin(\omega_{LO2}t) \\ = \; & -A/4\cos(\omega_{IF} - \omega_{LO2})t + B/4\cos(\omega_{IF} - \omega_{LO2})t \quad \text{(lowpassed)} \end{aligned} \tag{2.14}$$

Subtracting the signal in node D from node B, we obtain the output as $A/2\cos(\omega_{IF} - \omega_{LO2})t$ while the image term indicated by coefficient B are cancelled.

Same as in Hartley receiver, the gain mismatch and phase imbalance in Weaver receiver will also make the image cancellation not complete. Their effects on this architecture are also governed by (2.12). Note that all the mixers, lowpass filters and oscillators can produce gain and phase error.

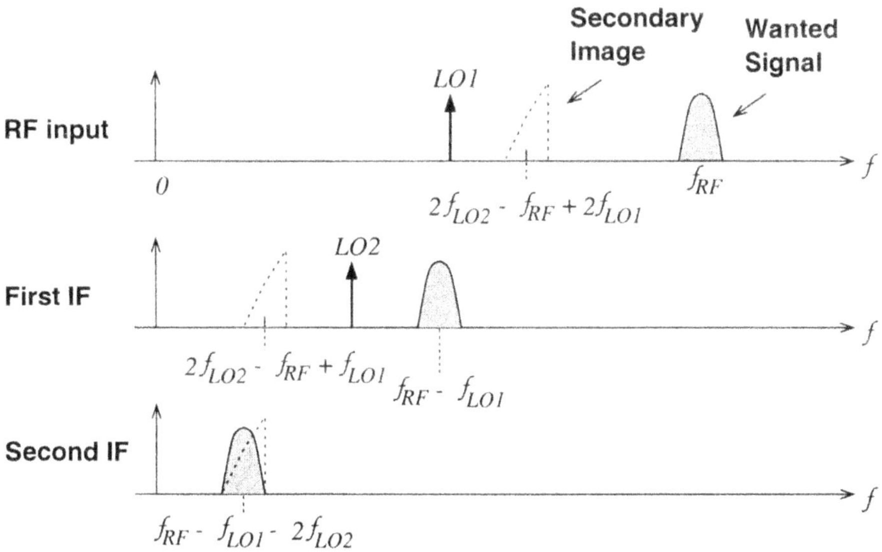

Figure 2.8: Secondary image problem in Weaver architecture.

If the receiver is configured with IF output, then the second mixing operation entails the problem of *secondary image*. To understand this issue, suppose the input spectrum contains an interferer at $2\omega_{LO2} - \omega_{RF} + 2\omega_{LO1}$. After the first downconversion, the interferer appears at $2\omega_{LO2} - \omega_{RF} + \omega_{LO1}$, that is, as the image of the signal with respect to ω_{LO2}. In the second downconversion, the interferer is not cancelled because it is originally on the same side of ω_{LO1} as the desired signal. Figure 2.8 illustrates this phenomenon. To suppress the secondary image, the lowpass filters in Figure 2.7 must be replaced with bandpass filters.

In Figure 2.7, if we sum up the signal at node B and D instead of subtracting them, then the image is left and the desired signal is eliminated. However, the second LO could be selected such that the image band is another desired signal band. This property can be utilised to build a *dual band receiver* [29]. A conceptual diagram is shown in Figure 2.9. By controlling the addition/subtraction operation, band 1 or band 2 can be selected. The advantage of this approach is obvious: all the hardware can be shared by the two band signal. The disadvantage is that the intermediate frequency, which must placed exactly at the center of the two bands, is fixed. There is no room for minimising the tuning range of the LO frequency synthesiser used in the whole transceiver [30], as usually done in a normal dual-band receiver [14].

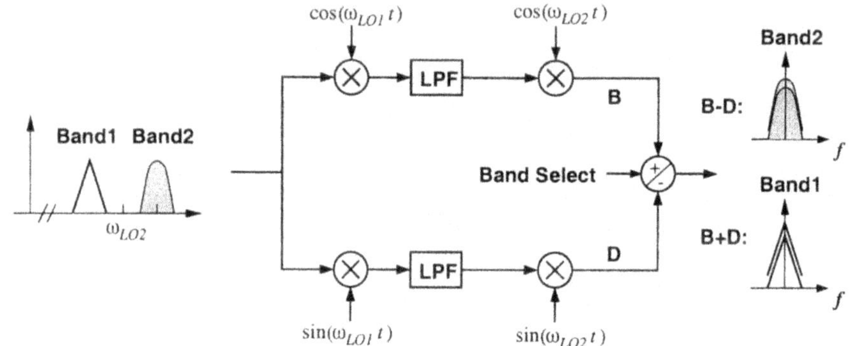

Figure 2.9: Dual-band implementation of the Weaver architecture.

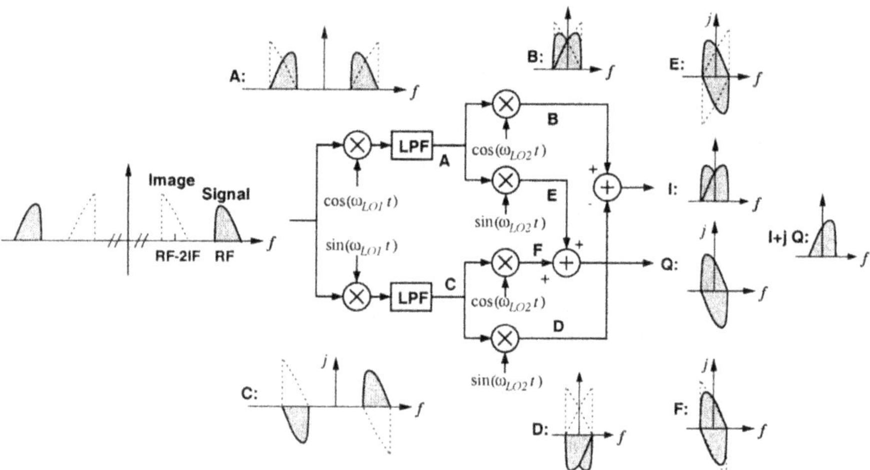

Figure 2.10: Weaver architecture with quadrature outputs.

The output of Figure 2.7 is a single channel, real signal. When quadrature outputs are required, a configuration shown in Figure 2.10 can be used. Another complex mixer is used for the IF-to-baseband conversion. This architecture is also referred as *complex IF receiver* as its IF is complex. Figure 2.10 also shows frequency spectra at different nodes of the receiver. The image interferer is cancelled out in the quadrature output by the same mechanism as the in-phase output. The final output spectrum, $I + jQ$, is asymmetric around zero frequency due to its complex nature. This means that the carried information of the received signal is doubled. This architecture has attracted much attention recently for its compatibility with today's demand of high-level of integration. In [31], an image rejection of 45 dB was reported.

2.5 Zero-IF Receiver

If the IF in a heterodyne receiver is reduced to zero, then the receiver is called as *zero-IF, homodyne*[1], or *direct conversion* receiver. Invented many decades ago, it has the simplest receiver topology.

Shown in Figure 2.11 is the block diagram of a zero-IF receiver with quadrature outputs. It consists of only an LNA, a I/Q mixer, two lowpass filters for anti-aliasing and channel selection, two A/D converters and two AGC units if required - almost the minimum set of circuit components required in any receiver. The simplicity of this receiver offers many advantages over the heterodyne receiver. Firstly and most importantly, image rejection is easy as the image is just the mirror of the signal itself (as explained later). Secondly, the LNA need not drive a 50 Ω load because no image rejection filter is required. Thirdly, the IF SAW filter and subsequent stages are replaced with lowpass filters and baseband amplifiers that are amenable to monolithic integration. It is for these reasons that this architecture has become a topic of active research presently [32, 33, 34, 35, 36, 37, 38, 39, 40, 41].

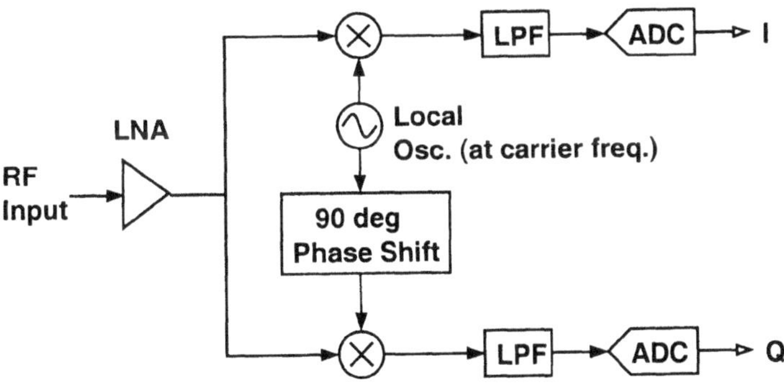

Figure 2.11: A zero-IF receiver with quadrature outputs.

[1] Historically, "homodyne" was restricted to the case that the local oscillator is synchronised in phase with the incoming carrier signal.

But, it has also significant drawbacks, The most important one is the DC offset produced by self-mixing of LO and interferers [3, 42], which often varies with time. Sophisticated DC offset cancellation is therefore required [43, 44, 45]. Other drawbacks include I/Q mismatch, even-order distortion, flicker noise and LO leakage.

Consider the image problem now. As the IF is now at DC, the image frequency is therefore $-f_{RF}$. On the other words, the image is the mirrored version of the desired signal about DC. The image is therefore called as *self-image*. As the complex (I/Q) mixer is used, the self-image is rejected. But due to the I/Q imbalances, the rejection is not complete and depends on how good the matching is. The mismatches include the quadrature LO phase errors and the gain and phase imbalances between any circuit components of the two paths.

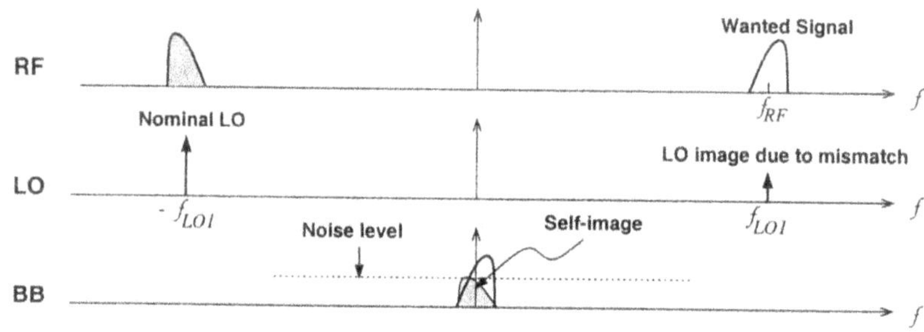

Figure 2.12: Self-image problem in a zero-IF receiver.

Figure 2.12 illustrates the self-image problem in the frequency domain. Without loss of generality, we assign all the gain and phase errors to the LO signal. Due to these errors, the complex LO signal is no longer a pure positive or negative frequency tone. A portion of the LO signal appears at the image frequency. To understand this issue, suppose the I and Q phases of LO are $(1+\alpha)\cos(\omega_{LO}t)$ and $-\sin(\omega_{LO}t+\varepsilon)$ respectively, where α and ε are gain and phase error respectively. The equivalent complex LO signal is:

$$\begin{aligned}
&(1+\alpha)\cos(\omega_{LO}t) - j\sin(\omega_{LO}t+\varepsilon) \\
&= \frac{1+\alpha-e^{j\varepsilon}}{2}e^{j\omega_{LO}t} + \frac{1+\alpha+e^{-j\varepsilon}}{2}e^{-j\omega_{LO}t} \\
&\approx \frac{(1+\alpha-\cos\varepsilon)-j\varepsilon}{2}e^{j\omega_{LO}t} + \frac{(1+\alpha+\cos\varepsilon)-j\varepsilon}{2}e^{-j\omega_{LO}t} \text{ for } \varepsilon \ll 1. \quad (2.15)
\end{aligned}$$

The coefficient of $e^{j\omega_{LO}t}$ in the above equation is not equal to zero, that is, the LO spectrum is not pure negative as shown in Figure 2.12. Multiplying with this LO, the downconverted signal consists of not only a nominal spectrum but also a mirrored version of itself, which actually is not much different than noise. Therefore, the signal-to-noise ratio is reduced.

To gain more insight to the self-image problem, suppose the received signal $x_{in} = a\cos\omega_c t + b\sin\omega_c t$, where a and b are either -1 or $+1$. Assume that the I and Q phases of the LO signal are $x_{LO,I}(t) = 2\cos\omega_c t$, $x_{LO,Q}(t) = 2(1+\alpha)\sin(\omega_C t + \varepsilon)$, where the factor two is included to simplify the results and α and ε represent gain and phase errors, respectively. Multiplying $x_{in}(t)$ by the two LO phases and lowpass filtering the result, we obtain the following baseband signals:

$$x_{BB,I} = a \tag{2.16}$$

$$x_{BB,Q} = (1+\alpha)b\cos\varepsilon - (1+\alpha)a\sin\varepsilon. \tag{2.17}$$

Therefore, the gain and phase error corrupt the downconverted signal constellation, thereby raising the bit error rate.

Figure 2.13(a) and (b) shows the resulting signal constellation with finite α or ε. This effect can be better seen by examining the downconverted signals in the time domain [Figure 2.13(c) and (d)]. Gain error simply appears as a non-unity scale factor in the amplitude. Phase imbalance, on the other hand, corrupts one channel with a fraction of the data pulses in the other channel, in essence degrading the signal-to-noise (SNR) ratio if the I and Q data streams are uncorrelated.

The problem of self-image is not serious in a single channel receiver with digital modulations which require typically only about 10 *dB* of SNR. For example, a 5° phase imbalance degrades the SNR requirements by roughly 1 *dB* only. However, it becomes very troublesome in a wideband receiver where the channel selection is carried out in digital domain, for example, in a multi-band AM receiver [46]. In this case, several radio channels are received, down-converted and digitised together. The "self-image" in this kind of receivers is actually an image from a neighbouring channel within the receiving band. Since very weak and very strong signals can appear at the same time, the image of a strong signal can be stronger than a weak signal. This makes the weak signal impossible to be retrieved. Figure 2.14 illustrates this problem.

2.6 Low-IF Receiver

To circumvent the problems of DC offset and $1/f$ noise in a zero-IF receiver and at the same time preserve most of its benefits, the IF can be translated to a low but nonzero value instead of to zero frequency. This kind of receiver is referred as *low-IF receiver*. Strictly speaking, any other types of receivers

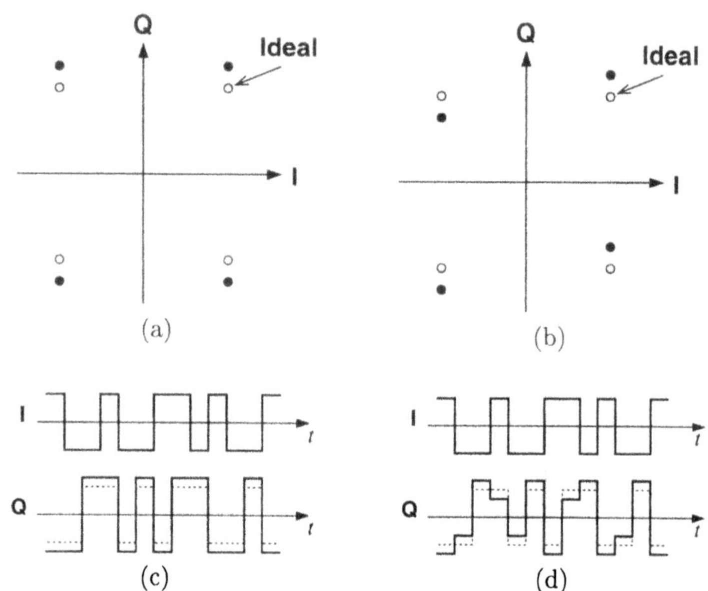

Figure 2.13: Self-image effect. Constellation (a) with gain error; (b) with phase error. Time-domain waveforms (c) with gain error; (d) with phase error.

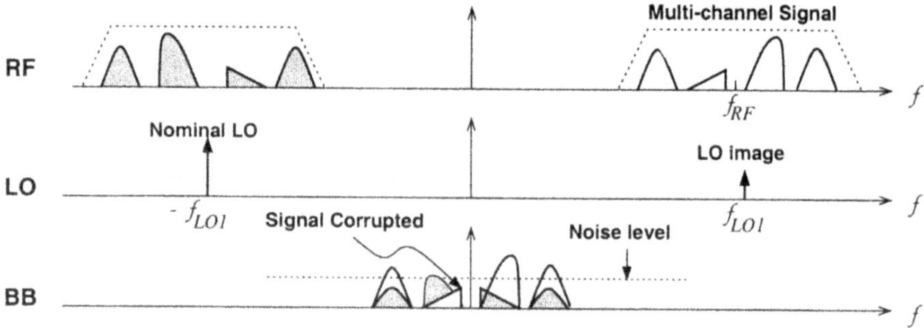

Figure 2.14: Self-image problem in a wideband zero-IF receiver.

can be categorised to this class as well if their IF is low. Due to low IF, normal monolithic filtering techniques such as $g_m - C$ or active RC continuous-time filters or SC filters can be still used for channel selection. Therefore, it offers both high performance and high degree of integration, and is therefore considered as a good candidate for realizing a fully integrated receiver.

To preserve the image rejection ability of the zero-IF receiver, a complex RF-to-IF mixer must be adopted in a low IF receiver. [47]. This requires a complex IF stage. Again, I/Q mismatches and LO phase error will limit the image rejection performance of this kind of receiver. A method for improving this performance is presented below.

2.6.1 Double Quadrature Downconverter

To improve the image rejection in the complex RF-to-IF mixer, an approach named *double quadrature downconversion* was proposed in [48]. Figure 2.15 shows a block diagram of the double quadrature downconverter. It employs two quadrature generators (90° phase shifter), one in LO path and another in the RF signal path, and two pairs of quadrature RF mixers that constitute an equivalent complex mixer. Note that all these components are integrable. The quadrature generator in the RF path is a passive asymmetric polyphase filter [49] which exhibits very high phase accuracy in a broad bandwidth. More details about polyphase filter are to be presented in Chapter four.

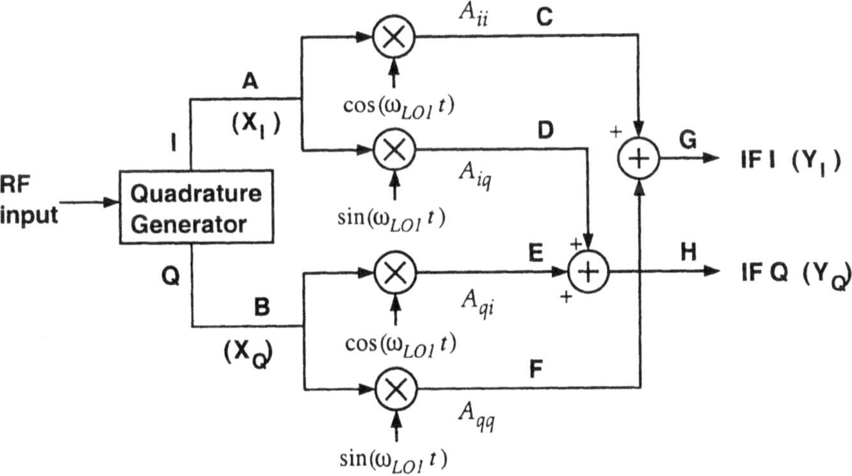

Figure 2.15: Low-IF receiver with double quadrature downconverter.

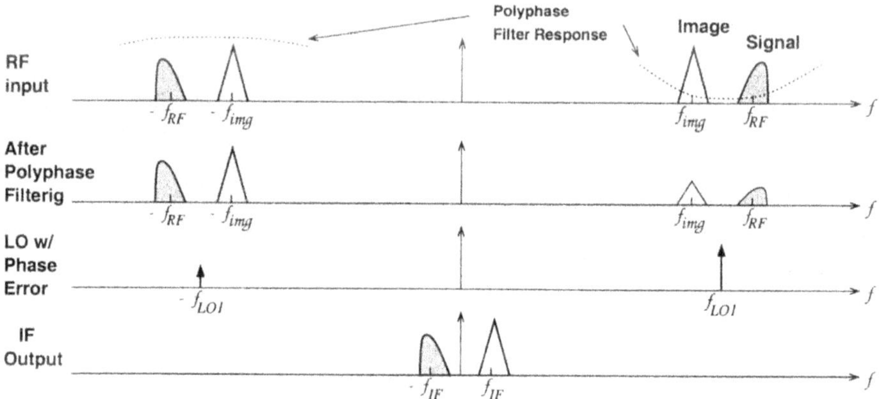

Figure 2.16: Spectra of the double quadrature downconversion.

Figure 2.16 shows the spectra of the double quadrature downconversion. The idea behind is that only the image interferer situated at positive frequency (if the complex LO is situated at positive axis) can be superimposed on the wanted signal after complex downconversion. This means that it is not necessary to suppress the image interferer at negative frequency axis, as is done with the classical high-Q RF filter. The suppression of only the negative frequency components does not require a high Q factor, even when the wanted and image frequency are situated very close to each other. The filtering can be done by a sequence asymmetric polyphase filter. The image is suppressed twice in a double quadrature downconverter: once by the polyphase filter and the other by the complex downconverter. So even with moderate phase accuracy of both LO and polyphase filter, very high image suppression can still be achieved. Obviously, the receiver has the drawback of using twice the number of RF mixers.

From another point of view, the double quadrature downconverter can be regarded as a combination of two Hartley image-reject downconverters: the first one is composed by loop A-B-C-F and gives in-phase output; the second one is composed by loop A-B-E-E and gives quadrature output. Phase error in the LO signal produces residual images in the I and Q output with the same magnitude. When we add jQ to I to obtain the complex output signal, these residual images will be cancelled out with each other.

Mismatch Analysis

Referring to Figure 2.15, the gain mismatch between path C and D or between path E and F will cause the cross coupling between I and Q branches of the IF output. This means that positive frequency signal will have a mirror signal in the negative frequency axis, and vice-versa. Since the wanted signal and the image are centered at -IF and +IF respectively, the cross coupling makes them to superimpose on each other.

To have an explicit expression of the mismatch effect, let us denote the mixer gain in the four paths as A_{ii}, A_{iq}, A_{qi} and A_{qq} as shown in Figure 2.15. We have the output $Y(j\omega)$ ($=Y_I(j\omega) + jY_Q(j\omega)$) in frequency domain as:

$$\begin{aligned} Y(j\omega) &= [A_{ii}LO_I(j\omega) \otimes X_I(j\omega) - A_{qq}LO_Q(j\omega)X_Q(j\omega)] \\ &\quad +j\,[A_{iq}LO_Q(j\omega) \otimes X_I(j\omega) + A_{qi}LO_I(j\omega) \otimes X_Q(j\omega)] \\ &= LO_{cm}(j\omega) \otimes [X_I(j\omega) + jX_Q(j\omega)] \\ &\quad +LO_{diff}(j\omega) \otimes [X_I(j\omega) - jX_Q(j\omega)], \end{aligned} \tag{2.18}$$

where $\otimes$ denotes convolution, $LO_I(j\omega)$ and $LO_Q(j\omega)$ are the Fourier transforms of I and Q LO signals respectively, $X_I(j\omega)$ and $X_Q(j\omega)$ are the Fourier transforms of I and Q input signals respectively, and

$$\begin{aligned} LO_{cm}(j\omega) = {} & \frac{A_{ii} + A_{qi} + A_{iq} + A_{qq}}{4}[LO_I(j\omega) + jLO_Q(j\omega)] + \\ & \frac{A_{ii} + A_{qi} - (A_{iq} + A_{qq})}{4}[LO_I(j\omega) - jLO_Q(j\omega)] \end{aligned} \tag{2.19}$$

$$\begin{aligned} LO_{diff}(j\omega) = {} & \frac{A_{ii} - A_{qi} + A_{iq} - A_{qq}}{4}[LO_I(j\omega) + jLO_Q(j\omega)] + \\ & \frac{A_{ii} - A_{qi} - (A_{iq} - A_{qq})}{4}[LO_I(j\omega) - jLO_Q(j\omega)]. \end{aligned} \tag{2.20}$$

Denote $X(j\omega) = X_I(j\omega)+jX_Q(j\omega)$. As x_I and x_Q are real signals, $X^*(-j\omega) = X_I(j\omega) - jX_Q(j\omega)$, where * represents complex conjugate. Therefore (2.18) becomes:

$$Y(j\omega) = LO_{cm}(j\omega) \otimes X(j\omega) + LO_{diff}(j\omega) \otimes X^*(-j\omega). \tag{2.21}$$

The output consists of two parts: input $X(j\omega)$ convolved with $LO_{cm}(j\omega)$ and the image of the input $X^*(-j\omega)$ convolved with $LO_{diff}(j\omega)$. The first part is the desired one and is shown in Figure 2.16. The second part is illustrated in Figure 2.17. It explains that the image interferer can be folded to the desired signal.

The conversion gains of the image and the desired signal are $(A_{ii} - A_{qi} - A_{iq} + A_{qq})/4$ and $(A_{ii} + A_{qi} + A_{iq} + A_{qq})/4$ respectively. The image rejection ratio is then given by

$$IRR = \frac{(A_{ii} + A_{qi} + A_{iq} + A_{qq})^2}{(A_{ii} - A_{qi} - A_{iq} + A_{qq})^2}. \tag{2.22}$$

Let A_{ii}, A_{qi}, A_{iq}, $A_{qq} = A \pm \Delta A$. The worst IRR is $(A/\Delta A)^2$. For example, for 1% gain mismatch among the four mixers, i.e., $\Delta A/A = 1\%$, the IRR of the double quadrature downconverter is limited to 40 dB.

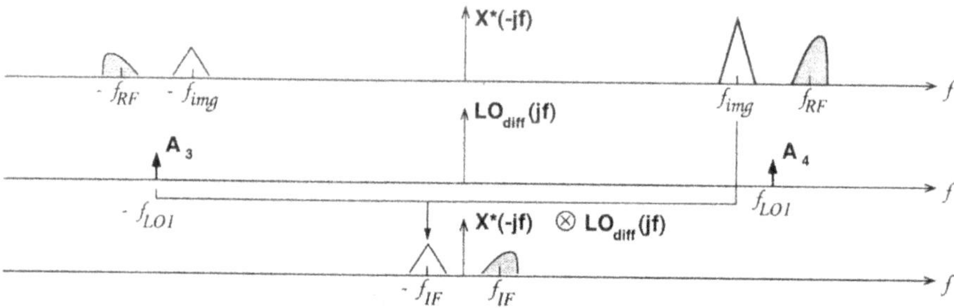

Figure 2.17: The image aliasing in a double quadrature downconverter, caused by gain mismatches. $A_3 = (A_{ii} - A_{qi} - A_{iq} + A_{qq})/4$ and $A_4 = (A_{ii} - A_{qi} + A_{iq} - A_{qq})/4$.

As a conclusion, the gain mismatches among the four mixers, instead of LO phase error, become the bottleneck for the double quadrature downconverter to achieve higher image rejection performance.

2.6.2 Direct-IF Digitising

As the IF is low in the low IF receiver, it is possible to perform the A/D conversion directly at IF stage. In such an implementation, the IF-to-baseband conversion and channel selection are carried out by digital circuitry. The direct-IF digitising has the advantage of perfect IF-to-baseband conversion and higher programmability. The cost is high power consumption in the data converter due as the high dynamic range requirement and high sampling frequency of the converter.

An efficient way of IF-to-baseband conversion is to set the sampling frequency of the ADC four times of the IF center frequency. This is a common practice in almost all IF ADCs [50, 51, 52, 53, 54]. With this sampling frequency, the digital LO signals become simply:

$$\text{LO I:} \qquad \cos\left(\frac{2\pi n}{4}\right) = 1, 0, -1, 0 \ldots, \qquad n = 0, 1, 2 \ldots \tag{2.23}$$

$$\text{LO Q:} \qquad \sin\left(\frac{2\pi n}{4}\right) = 0, 1, 0, -1 \ldots, \qquad n = 0, 1, 2 \ldots \tag{2.24}$$

No multiplication is required in the digital mixer.

Since the sampling frequency is very high in a direct IF digitising receiver, Nyquist-rate ADCs are difficult to achieve required dynamic range within reasonable power consumption. Oversampling ADCs with bandpass shaping of quantisation noise are usually adopted due to their high dynamic range. High speed $\Delta\Sigma$ ADC with IF up to 400 MHz has been reported with a resolution of 12-bit [55]. Bandpass $\Delta\Sigma$ ADCs [56] or lowpass $\Delta\Sigma$ ADCs [55, 57] with integrated IF mixer are common approaches. This topic will be further discussed in Chapter 5.

2.7 Software Radio

Recently, there is a trend to replace as much as possible analog functional blocks by digital signal processing units. In wireless receiver design, this appears as moving the ADC as close as possible to the RF front-end. The direct-IF digitising receiver mentioned earlier is in line with the trend. The ultimate destination of this trend is to locate the ADC just after the LNA as shown in Figure 2.18. This concept is known as *software radio*, or *software-defined radio* [58].

The software radio will cover all frequency bands, regardless of wireless standards. The functions of modulation/demodulation, up/down conversion, channel selection, etc., will be all implemented by real-time software. The output of the device can be voice, video, fax, data or any other forms of media. Therefore total flexibility will be provided. Also, it can easily adapt to new wireless standards without any hardware change.

Of our particular interest is that a software radio entails no image problem at all. The reason is obvious: there is only one signal path and no IF stage. Sources of the image problem, like non-zero IF, I/Q mismatch, LO phase errors and so on, are all circumvented.

All these nice features depend on the capability of the ADC. Software radio requires extremely large sampling rate, bandwidth and dynamic range in the ADC [59]. These difficulties make it unrealisable with today's technologies.

Nevertheless, people are going towards this direction. The presence of direct-IF digitising receiver is an example.

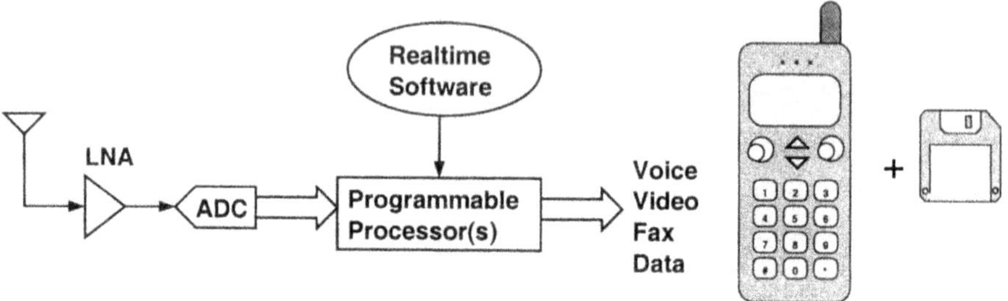

Figure 2.18: The concept of software radio.

2.8 Summary

In this chapter, the fundamental properties of the image problem associated with different receiver architectures have been presented. To understand the problem, the operational principles of various receiver architectures, including heterodyne, image-reject, zero-IF, low-If and direct-IF digitising, have been reviewed. Existing solutions to the image problem associated with them have been briefed.

It has been pointed out that the image problem is a major obstacle in achieving a fully integrated radio receiver. Hartley, Weaver or low IF receivers use complex mixers to relax or totally eliminate an image rejection filter. Practical implementation issues like finite matching condition, sufficient image suppression is very hard to achieve by the complex mixer. These problems demand innovative solutions in circuit or system levels.

References

[1] L. Lessing, *Man of High Fidelity: Edwin Howard Armstrong, a Biography*, New York: Bantam Books, 1969.

[2] Rolf Unbehauen and Andrzej Cichocki, *MOS Switched-Capacitor and Continuous-Time Integrated Circuits and Systems*, Springer-Verlag, 1989.

[3] A. A. Abidi, "Direct-conversion radio transceivers for digital communications," *IEEE J. Solid-State Circuits*, vol. 30, no. 12, pp. 1399–1410, Dec. 1995.

[4] V. Thomas et al., "A one-chip 2 GHz single-superhet receiver for 2Mb/s FSK radio communications," in *Digest of Technical Papers, IEEE Int. Solid-State Circuit Conference*, San Francisco, CA, Feb. 1994, pp. 42–43.

[5] T.D. Stetzler, I.G. Post, J.H. Havens, and M. Koyama, "A 2.7-4.5V single chip GSM transceiver RF integrated circuit," *IEEE J. Solid-State Circuits*, vol. 30, no. 12, pp. 1421–1429, Dec 1995.

[6] K. Irie, H. Matsui, T. Endo, et al., "A 2.7V GSM RF transceiver IC," in *Digest of Technical Papers, IEEE Int. Solid-State Circuit Conference*, Feb. 1997, pp. 302–303.

[7] P. Orsatti, F. Piazza, Q. Huang, and T. Morimoto, "A 20mA-receive 55mA-transmit GSM transceiver in 0.25μm CMOS," in *Digest of Technical Papers, IEEE Int. Solid-State Circuit Conference*, 1999, pp. 232–234.

[8] J. Schwartzel, "Filtering and frequency control for the next generation of mobile communication systems," in *Proc. EFTF*, 1994, pp. 175–189.

[9] S. Suma et al., "Surface mount type saw filter for hand-held telephones," in *Proc. Japan Electronic Manufacturing Technology Symposium*, June 1993, pp. 97–100.

[10] D. Grace and H. Iwatsubo, "RF filter technology for wireless communications," *Wireless Design & Development*, p. 8, June 1996.

[11] B.-S. Song and J. R. Barner, "A CMOS double-heterodyne FM receiver," *IEEE J. Solid-State Circuits*, vol. SC-21, no. 6, pp. 916–923, Dec. 1986.

[12] Francesco Piazza and Qiuting Huang, "A 1.57 GHz RF front-end for triple conversion GPS receiver," *IEEE J. Solid-State Circuits*, vol. 33, no. 2, pp. 202–209, Feb. 1998.

[13] M. Bopp, M. Alles, M. Arens, et al., "A DECT transceiver chip set using SiGe technology," in *Digest of Technical Papers, IEEE Int. Solid-State Circuit Conference*, 1999, pp. 69–69.

[14] Fordon J. Aspin, "RF system board level integration for mobile phones," in *Circuits and Systems for Wireless Communications*, M. Helfenstein and G. S. Moschytz, Eds., pp. 9–22. Kluwer Academic Publishers, 2000.

[15] J. Macedo, M. Copeland, and P. Schvan, "A 2.5-GHz monolithic silicon image filter," in *Proc. IEEE Custom Integrated Circuits Conference*, 1996, pp. 193–196.

[16] J. Macedo and M. Copeland, "A 1.9-GHz silicon receiver with monolithic image filtering," *IEEE J. Solid State Circuits*, , no. 3, pp. 378–386, Mar. 1998.

[17] Yuyu Chang and Jr. John Choma, "A monolithic RF image-reject filter," in *Proc. IEEE Southwest Symposium on Mixed Signal Design*, San Diego, California, USA, Feb 2000, pp. 41–44.

[18] Mitsubishi Microelectronics, *Mitsubishi Microwave/RF Datasheet MF1043S-1*, 1999.

[19] R.V.L. Hartley, "Modulation system," *U.S. Patent*, , no. 1 666 206, Apr. 17 1928.

[20] T. Okanobu, H. Tomiyama, and H. Arimoto, "Advanced low voltage single chip radio IC," *IEEE Trans. Consumer Electronics*, vol. 38, no. 3, pp. 465–475, August 1992.

[21] J. Tang and D. Kasperkovitz, "A 0.9-2.2GHz monolithic quadrature mixer oscillator for direct-conversion satellite receivers," in *Digest of Technical Papers, IEEE Int. Solid-State Circuit Conference*, Feb. 1997, pp. 88–89.

[22] M. Steyaert and R. Roovers, "A 1-GHz single-chip quadrature modulator," *IEEE J. Solid-State Ciruits*, vol. 27, no. 8, pp. 1194–1197, Aug 1992.

[23] Philips, *Philips RF/wireless communications data handbook*, Philips Semiconductor, 1996.

[24] John W. Archer, J. Granlund, and R. E. Mauzy, "A broad-band UHF mixer exhibiting high image rejection over a multidecade baseband frequency range," *IEEE J. Solid-State Circuits*, vol. SC-16, no. 4, pp. 385–392, Auguest 1981.

[25] M. D. McDonald, "A 2.5GHz BiCMOS image-reject front end," in *Digest of Technical Papers, IEEE Int. Solid-State Circuit Conference*, 1993, pp. 144–145.

[26] Werner Baumberger, "A single-chip image rejecting receiver for the 2.44 GHz band using commercial GaAs-MESFET-technology," *IEEE J. Solid-State Circuits*, vol. 29, no. 10, pp. 1244–1249, Oct. 1994.

[27] D. Pache, J.M. Fournier, G. Billiot, and P. Senn, "An improved 3V 2GHz BiCMOS image reject mixer IC," in *Proc. IEEE Custom Integrated Circuits Conference*, 1995, pp. 95–98.

[28] D.K. Weaver, "A third method of generation and detection of single-sideband signals," *Proc. IRE*, vol. 44, pp. 1703–1705, Dec 1956.

[29] Stephen Wu and Behzad Razavi, "A 900 MHz/1.8GHz CMOS receiver for dual-band applications," *IEEE J. Solid-State Circuits*, vol. 33, no. 12, pp. 2178–2185, Dec. 1998.

[30] Behzad Razavi, "Challenges in the design of frequency synthesizers for wireless applications," in *Proc. IEEE Custom Integrated Circutis Conference*, 1997, pp. 395–402.

[31] J.C. Rudell, J.J. Ou, et al., "A 1.9GHz wide-band IF double conversion CMOS receiver for cordless telephone application," *IEEE J. Solid-State Circuits*, vol. 32, pp. 2071–2088, Dec 1997.

[32] Y. Oishi, T. Takano, and H. Nakamura, "Sensitivity simulation results for a direct-conversion FSK receiver," in *Proc. IEEE Veh. technol. Conf.*, June 1988, pp. 588–595.

[33] K. Anvari, M. Kaube, and B. Hriskevich, "Performance of a direct conversion receiver with $\pi/4$-DQPSK modulated signal," in *Proc. IEEE Veh. technol. Conf.*, May 1991, pp. 822–827.

[34] J. Sevenhans et al., "An integrated Si bipolar RF transceiver for a zero IF 900 MHz GSM digital mobile radio frontend of a hand portable radio," in *Proc. IEEE Custom Integrated Circuits Conference*, May 1991, pp. 7.7.1–7.7.4.

[35] D. Haspeslagh et al., "BBTRX: A baseband transceiver for a zero IF GSM hand portable station," in *Proc. IEEE Custom Integrated Circuits Conference*, 1992, pp. 10.7.1–10.7.4.

[36] Christopher D. Hull, Joo Leong Tham, and Robert R. Chu, "A direct-conversion receiver for 900 MHz (ISM band) spread-spectrum digital cordless telephone," *IEEE J. Solid-State Circuits*, vol. 31, no. 12, pp. 1956–1963, Dec 1996.

[37] D. M. Binkley, J. M. Rochelle, B. K. Swann, et al., "A micropower CMOS, direct-conversion, VLF receiver chip for magnetic-field wireless applications," *IEEE J. Solid-State Circuits*, vol. 33, no. 3, pp. 344–358, Mar. 1998.

[38] S. Brett and G. Stanton, "A direct conversion L-band tuner for digital DBS," in *Digest of Technical Papers, IEEE Int. Solid-State Circuit Conference*, 1998, pp. 126–127.

[39] J. Itoh, M. Nishitsuji, O. Ishikawa, and D. Ueda, "2.1 GHz direct-conversion GaAs quadrature modulator IC for W-CDMA base station," in *Digest of Technical Papers, IEEE Int. Solid-State Circuit Conference*, 1999, pp. 226–227.

[40] A. P arssinen, J. Jussila, J. Ryyn anen, et al., "A wided-band direct conversion receiver for WCDMA applications," in *Digest of Technical Papers, IEEE Int. Solid-State Circuit Conference*, 1999, pp. 220–221.

[41] Thomas Cho, E. Dukatz, and M. Mack others, "A single-chip CMOS direct-conversion transceiver for 900 MHz spread-spectrum digital cordless phones," in *Digest of Technical Papers, IEEE Int. Solid-State Circuit Conference*, 1999, pp. 228–229.

[42] Behazad Razavi, "Design consideration for direct-conversion receivers," *IEEE Trans. on Circuits and Systems - II: Analog and Digital Signal Processing*, vol. 44, no. 6, pp. 428–435, June 1997.

[43] A. Bateman and D. M. Haines, "Direct conversion transceiver design for compact low-cost portable mobile radio terminals," in *Proc. IEEE Veh. technol. Conf.*, May 1989, pp. 57–62.

[44] S. Sampei and K. Fecher, "Adaptive DC-offset compensation algorithm for burst mode operated direct conversion receivers," in *Proc. IEEE Veh. Technol. Conf.*, 1992, pp. 93–96.

[45] J.K. Cavers and M.W. Liao, "Adaptive compensation for imbalance and offset losses in direct conversion transceivers," *IEEE Trans. Veh. Technol.*, pp. 581–588, Nov. 1993.

[46] E. van der Zwan, K. Philips, and C. Bastiaansen, "A 10.7MHz IF-to-basebad $\Delta\Sigma$ A/D conversion system for AM/FM radio receivers," in *Digest of Technical Papers, IEEE Int. Solid-State Circuit Conference*, Feb. 2000, pp. 340–341.

[47] A. Montalvo, A. Holden, W. Suter, et al., "A 22mW NADC receiver IF chip with integrated second IF channel filtering," in *Digest of Technical Papers, IEEE Int. Solid-State Circuit Conference*, 1999, pp. 222–223.

[48] J. Crols and M. Steyaert, "A 1.5GHz highly linear CMOS down conversion mixer," *IEEE J. Solid-State Circuits*, vol. 30, no. 7, pp. 736–742, July 1995.

[49] J. Crols and M. Steyaert, "An analog integrated polyphase filter for a high performance low-if receivers," in *Proc. VLSI Circuits Symposium*, Kyoto, June 1995, pp. 87–88.

[50] H.J. Dressler, "Interpolative bandpass A/D conversion - experimental results," *IEE Electron. Letters*, vol. 26, no. 20, pp. 1652–1653, Sept. 1990.

[51] A.M. Thurston, T.H. Pearce, and M.J. Hawksford, "Bandpass implementation of the sigma-delta A-D conversion technique," *Proc. IEE Int. Conference on A/D and D/A Conversion, Swansea, U.K.*, pp. 81–86, Sept. 1991.

[52] S.A. Jantzi, W.M. Snelgrove, and P.F. Ferguson Jr., "A fourth-order bandpass sigma-delta modulator," *IEEE J. Solid-State Circuits*, vol. 28, no. 3, pp. 282–291, March 1993.

[53] G. Tröster et al., "An interpolative bandpass converter on a 1.2μm BiCMOS analog/digital array," *IEEE J. Solid-State Circuits*, vol. 28, no. 4, pp. 471–477, April 1993.

[54] L. Longo and B.R. Horng, "A 15b 30kHz bandpass sigma-delta modulator," in *Digest of Technical Papers, IEEE Int. Solid-State Circuit Conference*, 1993, pp. 226–227.

[55] A. Namdar and B.H. Leung, "A 400-MHz, 12-bit, 18-mW, IF digitizer with mixer inside a sigma-delta modulator loop," *IEEE J. Solid-State Circuits*, vol. 34, no. 12, pp. 1765–1776, Dec. 1999.

[56] S.A. Jantzi, K.W. Martin, and A. S. Sedra, "Quadrature bandpass $\Delta\Sigma$ modulation for digital radio," *IEEE J. Solid-State Circuits*, vol. 32, no. 12, pp. 1935–1950, Dec. 1997.

[57] L.J. Breems, E.J. van der Zwan, and J.H. Huijsing, "A 1.8mW CMOS $\Sigma\Delta$ modulator with integrated mixer for a/d conversion of if signals," *IEEE J. Solid-State Circuits*, vol. 35, no. 4, pp. 468–475, April 2000.

[58] Joe Mitola, "Software radios," *IEEE Communications Magazine*, , no. 5, pp. 24–38, 1995.

[59] B. S. Song, "Low-spurious ADC architectures for software radio," in *Circuits and Systems for Wireless Communications*, M. Helfenstein and G. S. Moschytz, Eds., pp. 197–214. Kluwer Academic Publishers, 2000.

Chapter 3

Wideband 90° Phase Shifters

3.1 Introduction

The wideband 90° phase shifter is a critical building block in image-reject receivers, quadrature demodulators as described in the previous chapter, and many other applications where quadrature signal generation is required [1, 2, 3, 4, 5, 6]. The performance of the 90° phase shifter is directly linked to the image rejection performance of those receivers.

Mathematically, the 90° phase shifter is the *Hilbert transformer* [7], or a *Hilbert filter.* It is defined as a device in the form of a linear two-port whose output signal is a Hilbert transformation of the input signal. The strict definition of a Hilbert transformation is out of the scope of this book and can be found in many signal processing text books, for example in [8]. The transfer function of an ideal Hilbert transformer is given by:

$$H(j\omega) = \begin{cases} -j & \text{for} \quad \omega > 0 \\ 0 & \text{for} \quad \omega = 0 \\ j & \text{for} \quad \omega < 0. \end{cases} \tag{3.1}$$

The transfer function is illustrated in Figure 3.1. The magnitude $|H(j\omega)| = 1$ for all ω and the passband is infinite. The phase function is a step function: $\varphi(\omega) = -(\pi/2)sign(\omega)$, where function $sign(\omega)$ outputs the sign of ω. It shifts the phase of input signal by 90° from zero to infinite frequency.

Obviously, such an ideal Hilbert transformer is unrealisable in real world. Practical Hilbert transformers approximate the transfer function of (3.1) just

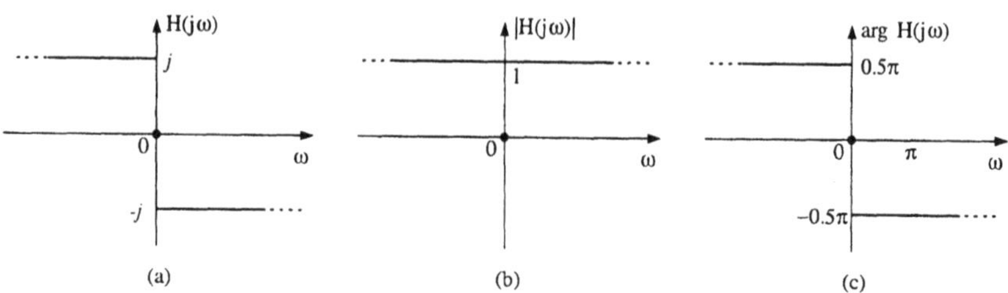

Figure 3.1: (a) The transfer function of an ideal Hilbert transformer, (b) its magnitude, and (c) its phase response.

in a certain frequency band, which is called the *care-band* of the transformer in this chapter. The accuracy of 90° phase-shift and care-band bandwidth are two most important performance parameters of a Hilbert transformer.

The Hilbert transformers can be implemented as an analogue or digital device. The focus of this chapter is on the analogue implementations, which can be in continuous-time (CT) or discrete-time (DT) domain.

There are many ways to implement CT Hilbert transformers, for example, delay lines, distributed couplers, lumped LC couplers and RC/CR allpass networks. However, only those circuits that are suitable for monolithic integration will be discussed in this chapter. These circuits are the conventional passive and active RC networks as well as asymmetric polyphase RC/CR networks [9]. Tuning circuit for these CT circuits is usually required because their edge frequencies depend on the absolute values of R and C. Otherwise the bandwidth of care-band must be over-designed.

The DT Hilbert transformers can be approximated by finite impulse response (FIR) and infinite impulse response (IIR) filters. Both of them can be realized by discrete-time circuit techniques, like switched-capacitor (SC) technique. In general, DT Hilbert transformers have two advantages over CT transformers: (1) wide effective bandwidth can be easily obtained; (2) edge frequency can be well controlled so that no tuning is required. However, they find applications in different systems. For example, if the system is indeed a DT system, then DT transformers should be used. Otherwise anti-aliasing or smoothing filters are required and this may increase the cost.

In this chapter, a conventional two-phase SC circuit realization and several new proposed SC realizations, including a polyphase circuit, a pseudo N-path circuit and a reduced opamp gain and bandwidth sensitivity circuit. All these circuits offer better performance and reduced complexity then the two-phase circuit. Some of the proposed SC Hilbert transformers will be used in the next two chapters.

This chapter is outlined as follows. First, several CT Hilbert transformers are reviewed. The design of FIR and IIR discrete-time Hilbert transformers is then described followed by switched-capacitor circuit realizations.

3.2 Continuous-Time Hilbert Transformers

As just mentioned, there are many well-established methods to implement a continuous-time analogue Hilbert transformer. However, only a few practical implementations which are suitable for monolithic integration will be discussed here. Other methods can be found in [10, 11, 12, 13, 14, 15, 16, 17, 18]

3.2.1 Passive RC/CR network

A pair of RC lowpass and CR highpass filter shown in Figure 3.2 consists of the simplest CT Hilbert transformer. The lowpass and highpass sections, which use same value for resistor R and capacitor C, have a same input port but different output ports.

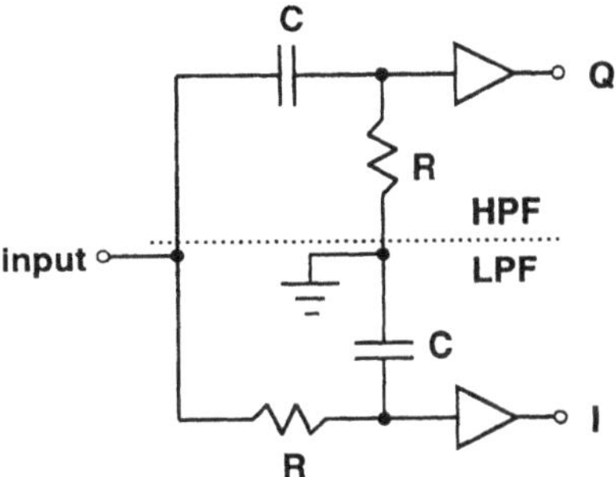

Figure 3.2: A RC/CR phase shifter.

The transfer functions of the lowpass section $H_{lp}(j\omega)$ and high pass section $H_{hp}(j\omega)$ are given by

$$H_{lp}(j\omega) = \frac{j\omega RC}{1+j\omega RC}, \quad \text{and} \tag{3.2}$$

$$H_{hp}(j\omega) = \frac{1}{1+j\omega RC}, \tag{3.3}$$

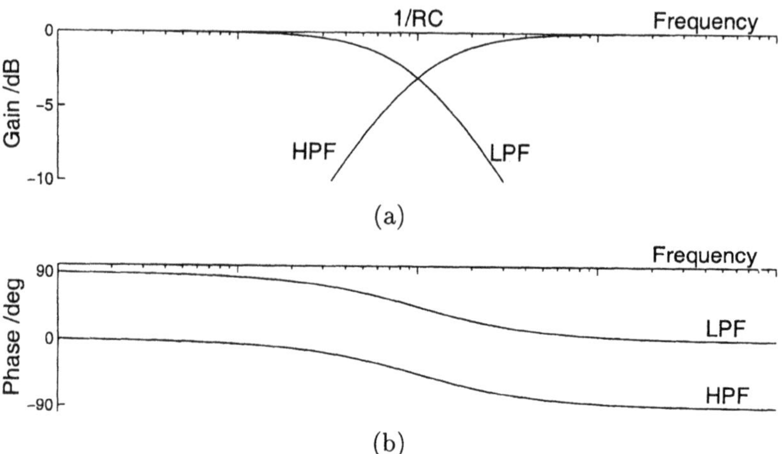

Figure 3.3: (a) Gain and (b) phase response of the LPF and HPF.

respectively. The gain and phase responses of the above equations are shown Figure 3.3. At $w = \pm 1/(RC)$, we have

$$\frac{H_{hp}(j\omega)}{H_{lp}(j\omega)} = -j\mathrm{sgn}(\omega), \tag{3.4}$$

that is, this pair of filters performs exactly the function of a Hilbert transformer at this frequency.

As shown in Figure 3.3(b), the RC lowpass section always lead the highpass section by 90° regardless of frequency. Practically, this phase leg may deviate from 90° because of the finite matching conditions of resistors and capacitors in integrated circuits.

To see how the matching accuracy determines the phase accuracy, suppose the values of resistor and capacitor are R and C respectively in the lowpass section, and are $R+\Delta R$ and $C+\Delta C$ respectively in the highpass section. The transfer function (3.3) of the highpass section becomes:

$$\begin{aligned} H_{hp}(j\omega) &= \frac{1}{1+j\omega(R+\Delta R)(C+\Delta C)} \\ &\cong \frac{1}{1+j\omega(RC+C\Delta R+R\Delta C)}. \end{aligned} \tag{3.5}$$

At $\omega = 1/RC$, the phase response are given by:

$$\angle H_{lp} = \pi/4, \tag{3.6}$$

$$\begin{aligned} \angle H_{hp} &= -\tan^{-1}\left(1 + \frac{\Delta C}{C} + \frac{\Delta R}{R}\right) \\ &\cong -\pi/4 + \frac{1}{2}\left(\frac{\Delta C}{C} + \frac{\Delta R}{R}\right). \end{aligned} \tag{3.7}$$

The last step above is obtained by the following approximation:

$$\tan(\pi/4 + \phi) \cong 1 + 2\phi, \quad \text{for } \phi \ll 1.$$

Therefore, the deviation of the phase difference between lowpass and highpass section from 90° is

$$-\frac{1}{2}\left(\frac{\Delta C}{C} + \frac{\Delta R}{R}\right)\frac{180}{\pi} \quad \text{degree.} \tag{3.8}$$

For example, suppose the matching conditions of capacitor and resistor are 1% and 2% respectively, we get 0.86^o phase error.

On the other hand, the magnitude responses of the highpass and lowpass section are matched at $\omega = 1/RC$ only. Note that this frequency depends on the absolute values of R and C that may have up to 30% variance in a typical IC technology. This is the main drawback of this CT Hilbert transformer. A solution to this problem is to employ resistors or capacitors which can be tuned externally [4].

Error correction techniques for improving the phase or amplitude matching accuracy exist in the literature [19, 9, 6]. Very impressive results were reported. For example, phase and gain errors as low as 0.1^o and $0.1dB$ respectively were reported in [9].

Due to its inherent low power consumption, the circuit of Figure 3.2 is widely used as the 90° phase shifter in Hartley image-reject receivers [20, 21] and quadrature modulators [4].

3.2.2 Asymmetric Polyphase Network

The problem of narrow amplitude matching band in the previous circuit can be solved in a sequence asymmetric polyphase RC networks [22] shown in Figure 3.4. This circuit has two input terminals which are connected to a differential input signal and four output terminals which give a pair of quadrature output signals in differential form. This polyphase network can be integrated

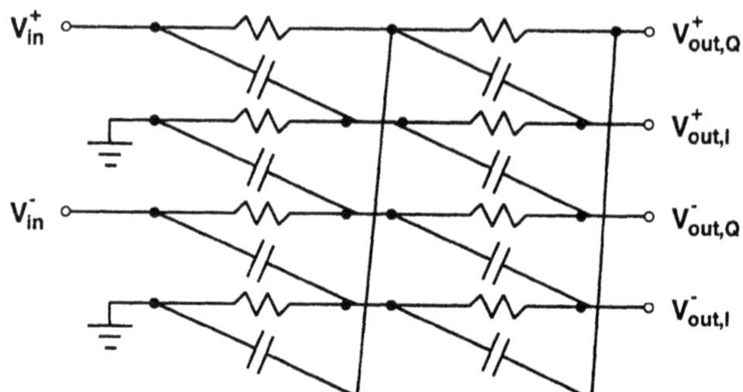

Figure 3.4: A two-stage sequence asymmetric RC polyphase network.

with reasonable resistor and capacitor values at frequency down to at least 10 MHz [23].

If the polyphase network has only one stage and only one input terminal is used with another grounded, then it works exactly the same as the previous circuit. The differences is that in a polyphase network several stages can be cascaded to increase its care-band bandwidth. The principle is described as follows. Firstly, as in the RC/CR network, the phase difference between the output signals of the polyphase network are 90° (if there is no component mismatches) and independent of frequency. Secondly, the amplitude error can be made small in a much wider bandwidth by using different values of R and C to obtain different center frequency in each stage. By over design the care-band, tuning or trimming process can be avoided. So the polyphase filter can endure large variations of R and C.

The explicit expression for the phase error with regard to component mismatch is very complex and hard to obtain since there are much more resistors and capacitors than a simple RC/CR network. But it can be estimated through Monte Carlo simulations.

Example 3.1. Suppose we want to design a polyphase network to generate the quadrature component of a LO signal at 1.7 GHz, and the variation of time constant RC are ±30% in the technology used. The center frequencies of two RC/CR stages can be designed as 2.2 GHz and 1.2 GHz respectively base on the the information of RC variance. The results are given as follows. Figure 3.5(a) shows the magnitude response of the equivalent complex output signal $v_{out,I}(j\omega)+jv_{vout,Q}(j\omega)$. From the figure we observe clearly two notches at -1.2 GHz and -2.2 GHz. Figure 3.5(b) shows the image rejection performance, i.e., the degree of suppression of image signal at negative frequency. From this

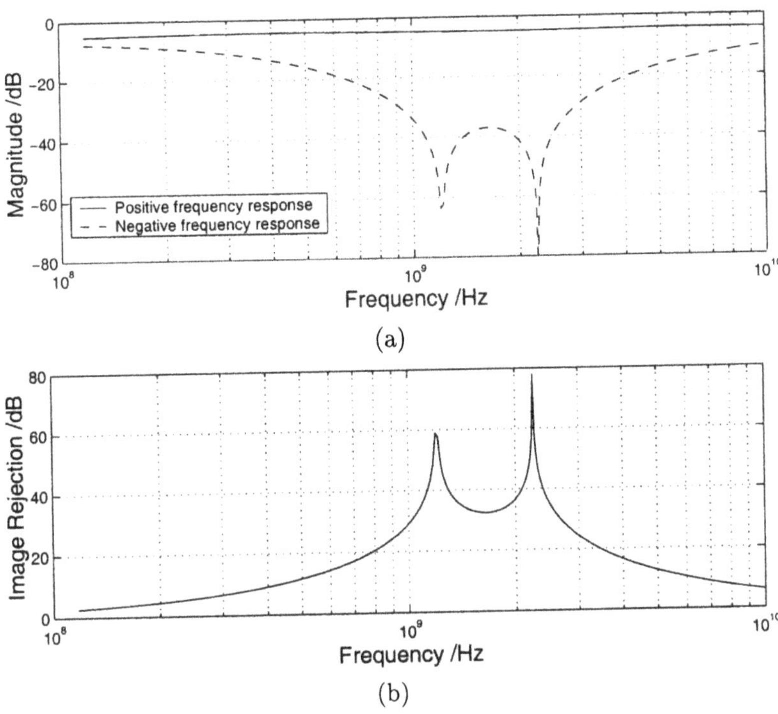

Figure 3.5: (a) Positive and negative frequency response and (b) image rejection response of a two-stage polyphase network.

figure, we can find that its image rejection ratio is better than 33 dB within the frequency range from 1.2 GHz to 2.2 GHz. The selection of R and C value for a given time constant RC depends on the input and output loading of the network. If buffers are inserted before the input and after the output, then the values of R or C can be arbitrary. However, these buffers dissipate large power in radio frequency range and are therefore not desirable.

□

By cascading more RC/CR stages, better image rejection can be achieved. However, without interstage buffers, signal will be attenuated due to the energy dissipation on resistors. Each stage adds 3 dB loss at its center frequency.

3.2.3 Active RC allpass network

The CT Hilbert transformer can be implemented by an allpass network in the form of a phase splitter as shown in Figure 3.6. It consists of two parallel allpass filters with a common input port and two separate output ports. The transfer functions of the allpass filters are:

$$H_1(j\omega) = e^{j\varphi_1(\omega)}; \quad H_2(j\omega) = e^{j\varphi_2(\omega)}. \tag{3.9}$$

The magnitude of both functions equals one. The phase difference of the signals at the output ports of the phase splitter should be

$$\delta(\omega) = \varphi_1(\omega) - \varphi_2(\omega) = -\pi/2; \quad \text{for all } \omega > 0. \tag{3.10}$$

The realization of this requirement is possible in a limited frequency band between the *low-frequency edge* ω_1 and the *high-frequency edge* ω_2. Note that in the above equation only positive-frequency part is considered since the phase functions are odd symmetric.

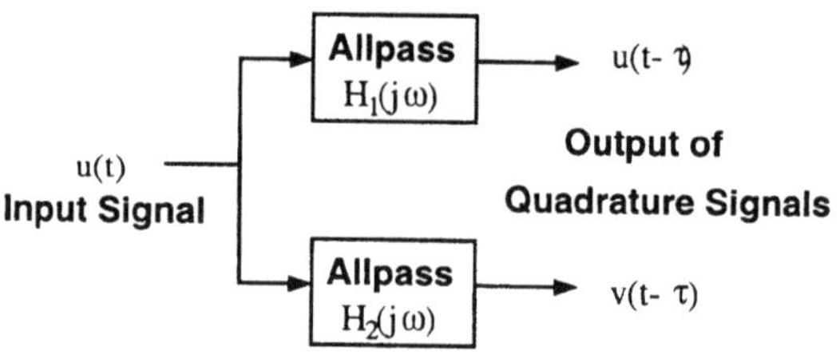

Figure 3.6: A Hilbert transformer in the form of phase-splitter.

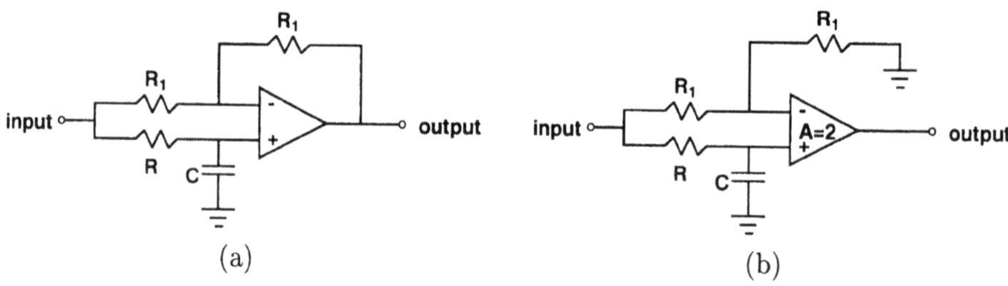

Figure 3.7: Active RC allpass circuits (a) using a normal opamp; (b) using a fixed gain amplifier.

Figure 3.7(a) shows an active RC allpass circuit [24, 25]. Assuming the opamp to be ideal, this circuit produces an allpass function of

$$H(j\omega) = -\frac{j\omega - \frac{1}{RC}}{j\omega + \frac{1}{RC}}, \tag{3.11}$$

which has a phase response of

$$\varphi(y) = \tan^{-1}\left[\frac{-2y}{1-y^2}\right]; \quad y = \omega RC. \tag{3.12}$$

The design of CT Hilbert transformers using allpass networks involves two steps. The first step is to linearise the phase function $\varphi_1(t)$ of $H_1(j\omega)$ in logarithm frequency domain. In the second step, the phase function of $H_w(j\omega)$ is obtained by shifting the function $\varphi(j\omega)$ in order to get a minimum value of the root-mean-square phase error of $\varphi_1(j\omega) - \varphi_2(j\omega) + \pi/2$ in frequency region $[\omega_1 \ \omega_2]$. The details of this method can be found in [7]. We just illustrate it with an example given below.

Example 3.2. In this example, a CT Hilbert transformer is implemented using two first-order allpass filters. The phase functions of the two allpass filters are

$$\varphi_1(j\omega) = \tan^{-1}\left[\frac{-2y}{1-y^2}\right]; \quad y = \omega\tau, \quad \tau = RC \tag{3.13}$$

$$\varphi_2(j\omega) = \tan^{-1}\left[\frac{-2ay}{1-a^2y^2}\right]. \tag{3.14}$$

For the normalised frequency edges $y_1 = 1.75$ and $y_2 = 3.5$ and the root-mean-square phase error of 0.012, the value of a is calculated to be 0.167.

Suppose we want to design a Hilbert transformer with $f_1 = \omega_1/2\pi = 10$ MHz. We have $f_2 = 20$ MHz. The required time constant of the allpass filters is given by equation $y_1 = 2\pi 10^7\tau = 1.75$. This yields $\tau = RC = 1.75/(2\pi 10^7) = 2.79 \times 10^{-8}$. There is a freedom in choosing R and C. The choice is dependent on input loading and parasitic resistance and capacitance. For example, let $C = 1$ pF, we have $R = 27.9$ $k\Omega$. The time constant of the second path equals $a\tau = 0.167 \times 2.7852 \times 10^{-8} = 4.65 \times 10^{-9}$. If $C = 1$ pF, we have $R = 4.65$ $k\Omega$. The function of $\varphi_1(j\omega)$ and $\varphi_2(j\omega)$ as well as the error function ϵ are shown in Figure 3.8(a) and (b) respectively.

□

The allpass filter shown in Figure 3.7(a) employs an opamp which has problems in high frequency operation. Recently, fixed gain amplifiers have been

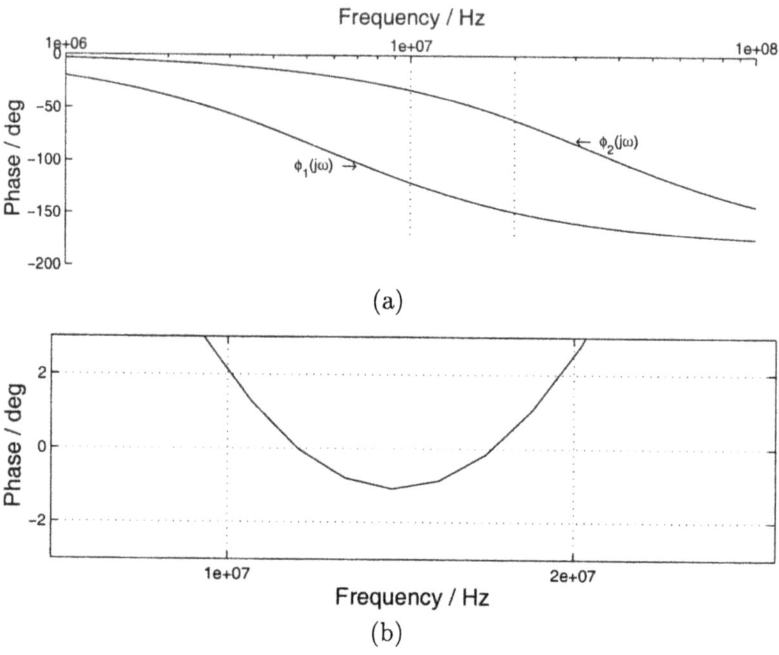

Figure 3.8: (a) The phase response and (b) the phase error of the CT Hilbert transformer implemented with two parallelled first order allpass filters with time constants of 2.79×10^{-8} and 4.65×10^{-9} respectively.

proposed as an alternative to opamps in creating amplifications where relative small fixed gains are required. These fixed gain amplifiers are integrable in a small area and are capable of very large bandwidths [26]. An allpass filler employing a fixed gain amplifier for realizing the transfer function $H(j\omega) = \left(j\omega - \frac{1}{RC}\right) / \left(j\omega + \frac{1}{RC}\right)$ is shown in Figure 3.7(b).

3.3 Discrete-Time Hilbert Transformers

The ideal discrete-time Hilbert transformer is defined as an allpass with a purely imaginary transfer function:

$$H(e^{j\psi}) = \begin{cases} -j & \text{for} \quad 0 < \psi < \pi \\ 0 & \psi = 0, \quad |\psi| = \pi \\ j & \text{for} \quad -\pi < \psi < 0 \end{cases} \tag{3.15}$$

where the variable $\psi = 2\pi f/f_s$ and f/f_s is a frequency normalised with respect to the sampling frequency f_s. The basic period has the interval from $-\pi$ to π. The transfer function is illustrated in Figure 3.9. The magnitude has the form

$$|H(e^{j\psi})| = \begin{cases} 1, & 0 < |\psi| < \pi \\ 0, & \psi = 0,; |\psi| = \pi \end{cases} \tag{3.16}$$

and the phase function can be written as

$$\arg[H(e^{j\psi})] = -(\pi/2)\mathrm{sgn}[\sin(\psi)]. \tag{3.17}$$

Besides, the magnitude of $1 + jH(z)$ equals two for $0 < \psi < \pi$, 0 for $-\pi < |\psi| < 0$ and 1 for $|\psi| = \pi$ as shown in Figure 3.9(d). By expanding the transfer function (3.15) into an infinitive Fourier series and performing inverse Fourier transform on it, then the corresponding ideal impulse response is obtained as:

$$h(n) = \begin{cases} \frac{\pi}{2}\frac{\sin^2(\pi n/2)}{n} & n \neq 0 \\ 0 & n = 0. \end{cases} \tag{3.18}$$

Once again, such an ideal discrete-time Hilbert transformer is noncausal and unrealisable. It is possible to approximate the ideal frequency response in (3.15) by two methods. The first method uses a pair of phase splitters, whose outputs, with a common input, bear a Hilbert transform relationship with each other. This is illustrated in Figure 3.10, where the two channels are called *filter-I* and *filter-Q* and have transfer functions $H_I(e^{j\psi})$ and $H_Q(e^{j\psi})$ respectively. The transformer can be fully characterised by its complex transfer function $H_I(e^{j\psi}) + jH_Q(e^{j\psi})$, which tends to zero at $-\pi < \psi < 0$. The second method uses a single causal filter whose output approximates the Hilbert transform of a delayed version of the input signal. We will focus on the first method and the corresponding FIR and IIR design approaches are described in the following paragraphs.

3.3.1 Design of FIR Hilbert Transformers

There are a number of methods to design a discrete-time FIR Hilbert transformer. For example, by truncating with a window function and shifting of the ideal impulse response (3.18), the coefficients of a causal FIR Hilbert transformer are obtained. The FIR Hilbert transformer can be also derived from corresponding designs for symmetric half-band filters [27, 28]. In this section, the design of a specially useful FIR Hilbert transformer which has the care-band centered at $\psi = -\pi/2$ is introduced.

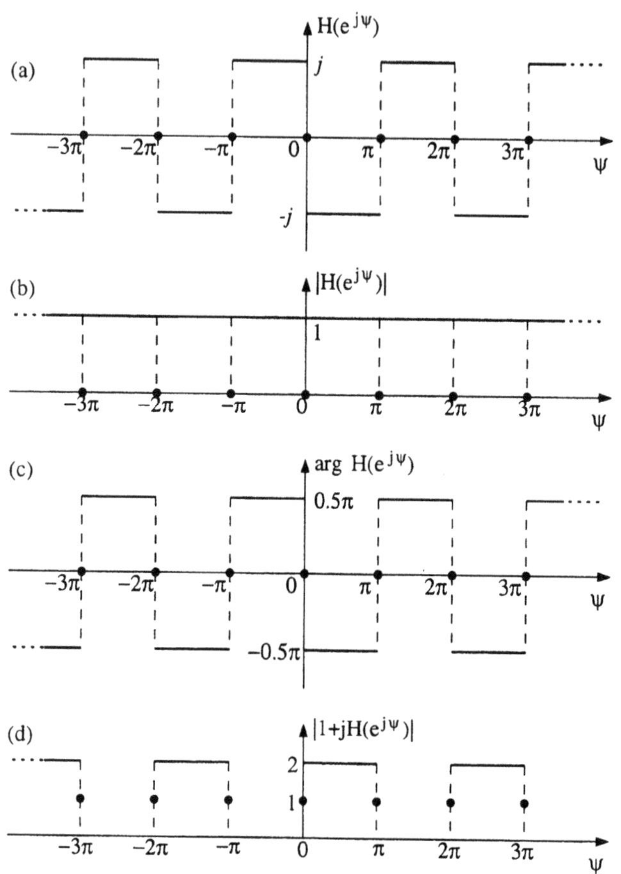

Figure 3.9: (a) The transfer function of an ideal discrete-time Hilbert transformer $H(z)$, (b) its magnitude, (c) its phase response and (d) the magnitude of $1 + jH(z)$.

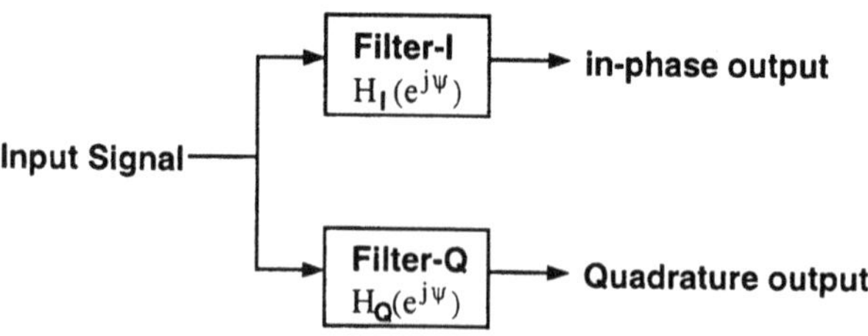

Figure 3.10: A discrete-time version of the phase-splitter Hilbert transformer.

The FIR Hilbert transformer can be designed by assigning zeros to its complex transfer function $H(z) = H_I(z) + jH_Q(z)$ in the region of $-\pi < \psi < 0$. A zero in the transfer function of a causal FIR filter is realized by the following factor:

$$1 - z^{-1}e^{-j\psi_1}, \quad (z = e^{j\psi}, \quad \psi = 2\pi f/f_s) \tag{3.19}$$

where the zero is located at $\psi = \psi_1$. In our approach to design an FIR Hilbert transformer, zeros are allocated at $\psi = -\frac{\pi}{2}$, or appear as a symmetric pair around $-\frac{\pi}{2}$, i.e., $\psi = -\frac{\pi}{2} \pm \phi$. Therefore, the general form of the transfer function is:

$$\begin{aligned} H(z) &= (1 - z^{-1}e^{-j\frac{\pi}{2}})^k \prod_{i=1}^{m}(1 - z^{-1}e^{-j(\frac{\pi}{2}+\phi_i)})(1 - z^{-1}e^{-j(\frac{\pi}{2}-\phi_i)}) \\ &= (1 + jz^{-1})^k \prod_{i=1}^{m}(1 + j2\cos\phi_i z^{-1} - z^{-2}), \end{aligned} \tag{3.20}$$

where k is the number of zeros located at $\psi = -\frac{\pi}{2}$ and m is the number of zero pairs around at $-\frac{\pi}{2}$. Expanding the above equation and separating terms with real and imaginary coefficients, the transfer functions $H_I(z)$ and $H_Q(z)$ are obtained.

If the FIR Hilbert transformer has odd number of zeros at $\psi = -\pi/2$, then the transfer function $H_I(z)$ and $H_Q(z)$ can be expressed in the form [29]:

$$H_I(z) = a_0 - a_1 z^{-2} + a_2 z^{-4} + \ldots + (-1)^n a_n z^{-2n}, \tag{3.21}$$

$$H_Q(z) = a_n z^{-1} - a_{n-1} z^{-3} + a_{n-2} z^{-5} + \ldots + (-1)^n a_0 z^{-(2n+1)}, \tag{3.22}$$

where n is a non-negative integer. The order of the FIR Hilbert transformer is defined as the highest power of z^{-1}, i.e., $2n + 1$. The first coefficient a_0 is usually normalised to 1. The $H_I(z)$ and $H_Q(z)$ have same magnitude at all frequencies. This is proven as follows. Since

$$\begin{aligned} H_Q(z) &= a_n z^{-1} - a_{n-1} z^{-3} + a_{n-2} z^{-5} + \ldots + (-1)^n a_0 z^{-(2n+1)} \\ &= (-1)^n z^{-(2n+1)}(a_0 - a_1 z^2 + a_2 z^4 + \ldots + (-1)^n a_n z^{2n}) \\ &= (-1)^n z^{-(2n+1)}(a_0 - a_1 z^{-2} + a_2 z^{-4} + \ldots + (-1)^n a_n z^{-2n})^* \\ &= (-1)^n z^{-(2n+1)} H_I^*(z), \end{aligned} \tag{3.23}$$

where * denotes complex conjugate. Therefore we have

$$|H_I(e^{j\psi})| = |(-1)^n e^{-j(2n+1)\psi}|\,|H_I^*(e^{j\psi})| = |H_I(e^{j\psi})|$$

for all ψ.

Example 3.3. If only one zero at $\psi = -\pi/2$ is assigned to the transfer function, then we obtain the simplest FIR Hilbert transformer which has

$$\begin{cases} H_I(z) = 1, \\ H_Q(z) = z^{-1}. \end{cases} \tag{3.24}$$

The magnitude of the complex transfer function $H_I(z) + jH_Q(z)$ and the corresponding pole-zero plot are shown in Figure 3.11 (a) and (b) respectively, denoted as "FIR1". Phase responses of $H_I(z)$ and $H_Q(z)$ and their difference are given in Figure 3.12(a) and (e) respectively. These figures shows that this transformer approximates the ideal transformer only in a small vicinity of $\psi = -\pi/2$. However, due to its simplicity, this transformer will be used in a complex IF oversampling $\Delta\Sigma$ converter to perform a double image rejection, to be described in Chapter five.

□

Example 3.4. If three zeros at $\psi = -\pi/2$ are assigned to the transfer function $H(z) = H_I(z) + jH_Q(z)$, we obtain

$$\begin{aligned} H(z) &= (1 + jz^{-1})^3 \\ &= 1 - 3z^{-2} + j(3z^{-1} - z^{-3}). \end{aligned} \tag{3.25}$$

Therefore,

$$\begin{cases} H_I(z) = 1 - 3z^{-2}, \\ H_Q(z) = 3z^{-1} - z^{-3}. \end{cases} \tag{3.26}$$

The magnitude of the complex transfer function $H_I(z) + jH_Q(z)$ and the corresponding pole-zero plot are shown in Figure 3.11 (a) and (b) respectively, denoted as "FIR2". Phase responses of $H_I(z)$ and $H_Q(z)$ and their difference are given in Figure 3.12(b) and (f) respectively. From Figure 3.11(a), we find that the stop-band region is centered at $-f_s/4$ with a bandwidth of $0.07f_s$, and the stop-band attenuation is more than 60 dB. The coefficients of (3.26) are all integers and have small spread in value, thus it is suitable for IC implementation.

□

Example 3.5. If one zero at $\psi = -\pi/2$ and two zeros at $\psi = -\pi/2 \pm \phi$ are assigned to the transfer function $H(z) = H_I(z) + jH_Q(z)$, we obtain

$$\begin{aligned} H(z) &= (1 + jz^{-1})(1 + j2\cos\phi z^{-1} - z^{-2}) \\ &= 1 - (1 + 2\cos\phi)z^{-2} + j[(1 + 2\cos\phi)z^{-1} - z^{-3}]. \end{aligned} \tag{3.27}$$

Therefore,

$$\begin{cases} H_I(z) = 1 - (1 + 2\cos\phi)z^{-2}, \\ H_Q(z) = (1 + 2\cos\phi)z^{-1} - z^{-3}. \end{cases} \tag{3.28}$$

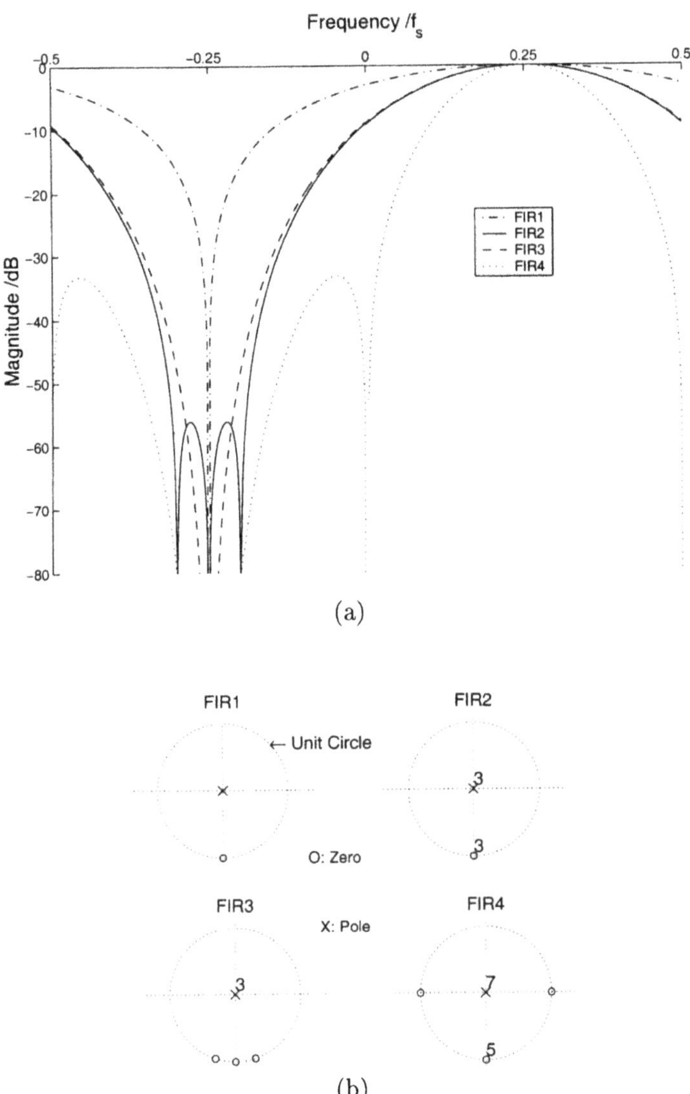

Figure 3.11: (a) Magnitude response of the complex transfer function $H_I(z) + jH_Q(z)$ of the four FIR Hilbert transformer examples. (b) The corresponding pole-zero plot.

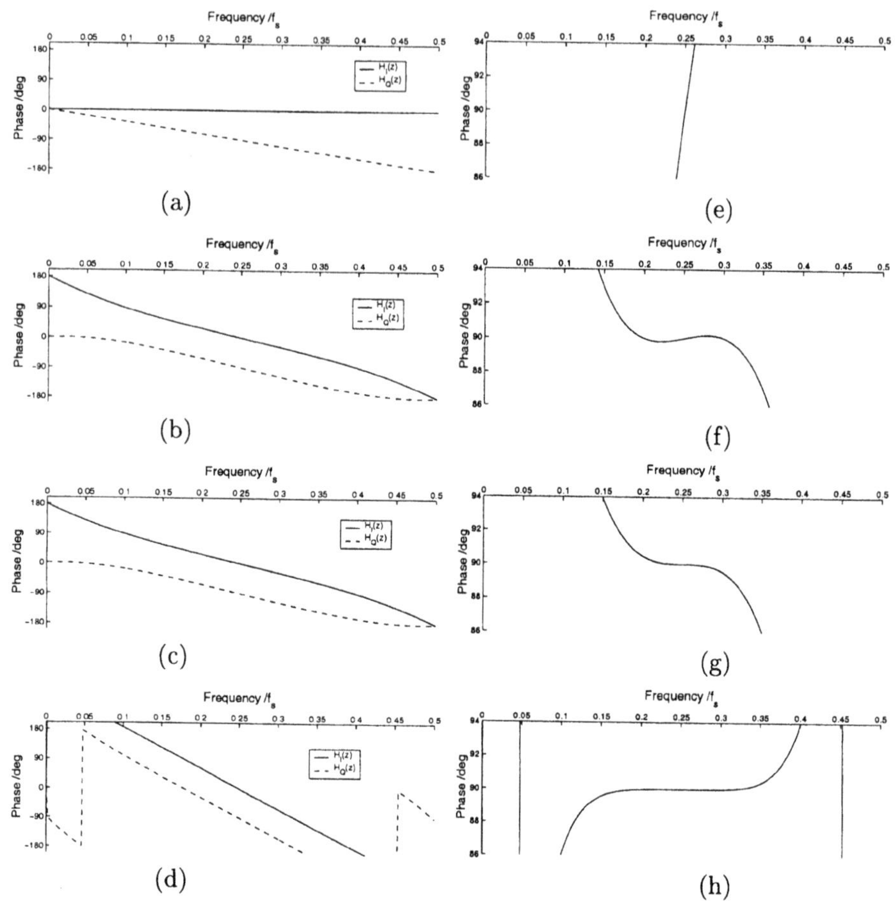

Figure 3.12: (a)-(d) Phase responses of $H_I(z)$ and $H_Q(z)$, and (e)-(h) phase differences between $H_I(z)$ and $H_Q(z)$ of the four FIR Hilbert transformer examples.

There are two peaks located in-between three zeros at $-\pi/2-\phi, -\pi/2, -\pi/2+\phi$. By solving

$$\frac{\partial |H_I(e^{j\psi}) + jH_Q(e^{j\psi})|}{\partial \psi} = 0 \tag{3.29}$$

the peak location ψ_p is found as:

$$\psi_p = -\cos^{-1}\left(\pm\frac{1}{6}\sqrt{27-6a_1-a_1^2}\right) \tag{3.30}$$

where $a_1 = 1 + 2\cos\phi$. The corresponding peak R_s is

$$R_s = \frac{(3-a_1)^2}{27(1+a_1)^2}. \tag{3.31}$$

Substituting $a_1 = 1 + 2\cos\phi$ to (3.31), the relation between R_s and ϕ is found:

$$R_s = \frac{4}{27}\frac{\sin^6\phi/2}{\cos^2\phi/2}. \tag{3.32}$$

Therefore, a trade-off between the bandwidth ϕ and stop band peak R_s exists: the wider is the bandwidth, the higher is the stop band peak, and vice versa.

A numerical example is given here. Let $\phi = \frac{\pi}{10}$, which implies a bandwidth of $2\phi = 2\pi/10$. Since $a_1 = 1 + 2\cos\phi$, we have the filter coefficient $a_1 = 2.9$ and the transfer function is

$$\begin{cases} H_I(z) = 1 - 2.9z^{-2}, \\ H_Q(z) = 2.9z^{-1} - z^{-3}. \end{cases} \tag{3.33}$$

From (3.31), the stop band peak is $-56dB$. Its magnitude response and pole-zero plot are shown in Figure 3.11(a) and (b) respectively, denoted as "FIR3". Phase responses of $H_I(z)$ and $H_Q(z)$ and their difference are given in Figure 3.12(c) and (g) respectively.

□

Example 3.6. If five zeros at $\psi = -\pi/2$ and two at $\psi = 0$, π are assigned to the FIR Hilbert transformer, then its transfer function is:

$$\begin{cases} H_I(z) = 1 - 11z^{-2} + 15z^{-4} - 5z^{-6} \\ H_Q(z) = 5z^{-1} - 15z^{-3} + 11z^{-5} - z^{-7}. \end{cases} \tag{3.34}$$

Its magnitude response and pole-zero plot are shown in Figure 3.11 (a) and (b) respectively, denoted as "FIR4". Phase responses of $H_I(z)$ and $H_Q(z)$ and their difference are shown in Figure 3.12(d) and (h) respectively. Note that all coefficients of this transformer are integers. This makes it suitable for IC implementation. This transformer was actually used in a special I/Q demodulator [29].

□

3.3.2 Design of IIR Hilbert Transformers

The IIR Hilbert transformer design method described here was introduced by Ansari in [30]. This kind of IIR Hilbert transformers are implemented by use of noncausal generalised half-band filters which are derived by modifying the conventional elliptic filter design so that all poles of the half-band filter lie on the imaginary axis. The ideal IIR half-band transfer function has the form [31]

$$H_{HB}(z) = 1 + z^{-1}G(z^2) \tag{3.35}$$

where $G(z^2)$ is an allpass filter with unit magnitude. The $H_{HB}(z)$ is ideally equal to $+2$ in the passband and 0 in the stop band as shown in Figure 3.13(a). Let us show step by step, as displayed in Figure 3.13(a-e), that the transfer function of an ideal IIR Hilbert transformer is given by

$$H(z) = z^{-1}G(-z^2). \tag{3.36}$$

We denote

$$F(z) = z^{-1}G(z^2) \tag{3.37}$$

which has a unit magnitude and a phase function equal to zero in the passband and $\pm\pi$ in the stopband, as shown in Figure 3.13(c). This phase function can be written in the form

$$\Phi(\psi) = 0.5\pi\{\text{sgn}[\sin(2\psi)] - \text{sgn}\psi\}. \tag{3.38}$$

If we now multiply $F(z)$ by j, we get

$$jF(z) = je^{j\Phi(\psi)} = e^{j[\Phi(\psi)+0.5\pi]}, \tag{3.39}$$

i.e., the phase function is lifted by 0.5π as shown in Figure 3.13(d). By replacing z by jz in $jF(z)$, the phase function is then shifted along the ψ axis by 0.5π and yields the end result shown in Figure 3.13(e); that is, the phase function of the ideal Hilbert transformer given by (3.17). Note that all the transformations do not change the amplitude function of the $F(z)$. Therefore,

$$H(z) = j(jz)^{-1}G[(-jz)^2)] = z^{-1}G(-z^2). \tag{3.40}$$

The Hilbert transformer can be designed to have equiripple phase function and exactly flat amplitude distribution using generalised half-band elliptic filters. As proposed in [31], the $G(z^2)$ can be expressed as

$$G(z^2) = z^2\frac{H_1(z^2)}{H_0(z^2)} \tag{3.41}$$

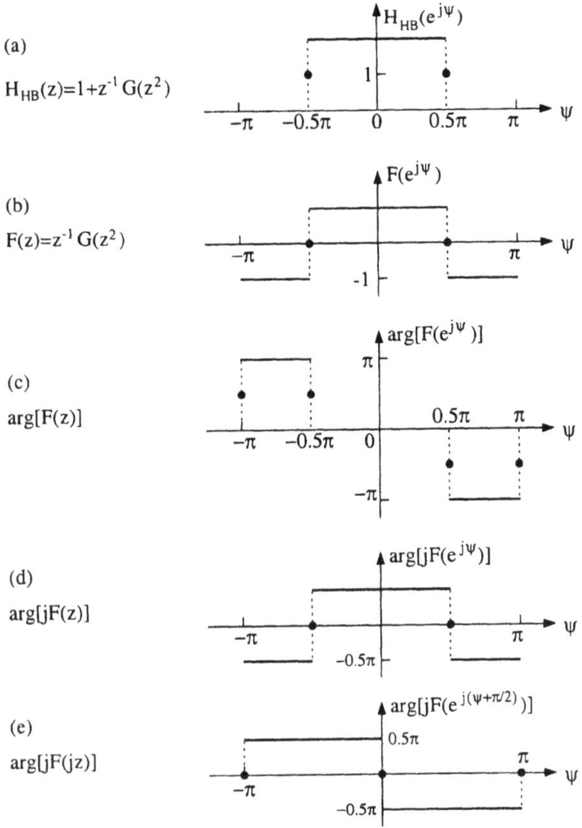

Figure 3.13: Step-by-step derivation of the IIR transfer function of a Hilbert transformer defined by (3.36), starting from the transfer function of the ideal half-band filter given by (3.35).

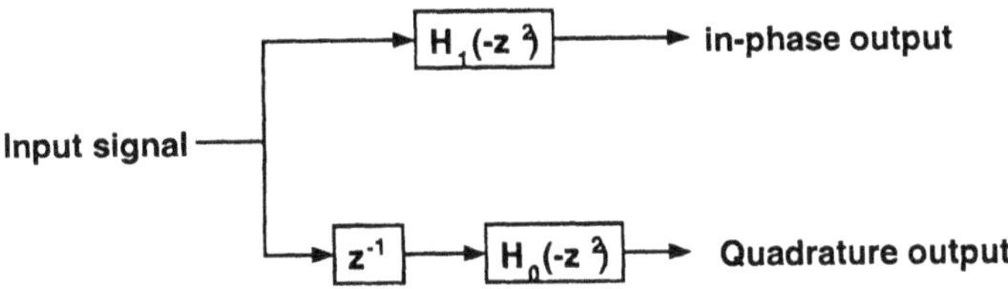

Figure 3.14: Filter $H_1(-z^2)$ and $z^{-1}H_0(-z^2)$ constitute a pair of Hilbert transformer.

where $H_1(z^2)$ and $H_0(z^2)$ are allpass with poles inside the unit circle. Therefore, $H_1(-z^2)$ and $z^{-1}H_0(-z^2)$ constitute a pair of 90° phase shifter as shown in Figure 3.14. The explicit form of the transfer function of the Hilbert transformer is given by

$$H(z) = z^{-1}\prod_{i=1}^{N}\frac{z^{-2}-a_i}{1-a_iz^{-2}}; \quad z = e^{j\psi} \tag{3.42}$$

where N is a positive integer and defined as the order of the transformer, a_i's are filter coefficients. The magnitude of this transfer function equals exactly one for all ψ. Therefore, the passband of this Hilbert transformer is uniquely defined by its phase function. The calculation of coefficients a_i is quite involving. A Matlab program for obtaining these coefficients is given in Appendix 3.A. Some a_i's may be greater than unity. To avoid being unstable, filter $H_0(z)$ and $H_1(z)$ are chosen as follows. Let I_0 be the set of integers that $a_i > 1$ and $1 \leq i \leq L$. Also let $I_1 = I_0^c \cap \{0, 1, \cdots, L\}$. The $H_0(-z^2)$ are given by

$$H_0(-z^2) = \prod_{i\in I_0}\frac{a_iz^{-2}-1}{z^{-2}-a_i}, \quad a_i > 1 \tag{3.43}$$

and $H_1(-z^2)$ are given by

$$H_1(-z^2) = \prod_{i\in I_1} -\frac{z^{-2}-a_i}{1-a_iz^{-2}}, \quad 0 < a_i < 1, \tag{3.44}$$

so that both $H_0(-z^2)$ and $H_1(-z^2)$ have poles inside unit circle, and therefore be stable. Several examples of IIR Hilbert transformers derived from generalised half-band filter are illustrated below.

Example 3.7. Consider the passband and stop-band width of 0.15π, and passband and stop-band ripple of less than 0.01 and 0.001 respectively. With the Matlab program, we get a transfer function with $N = 1$ and coefficients $a_1 = 0.338019$. Therefore, the corresponding filter-I and filter-Q of the IIR Hilbert transformer are given by:

$$H_I(z) = z^{-1}, \tag{3.45}$$

$$H_Q(z) = \frac{z^{-2}-0.338019}{1-0.338019z^{-2}}. \tag{3.46}$$

Its magnitude response ($|H_I(z) + jH_Q(z)|$) and phase response ($\angle H_I(z) - \angle H_Q(z)$) are shown in Figure 3.15. From the figure, we find that the actual

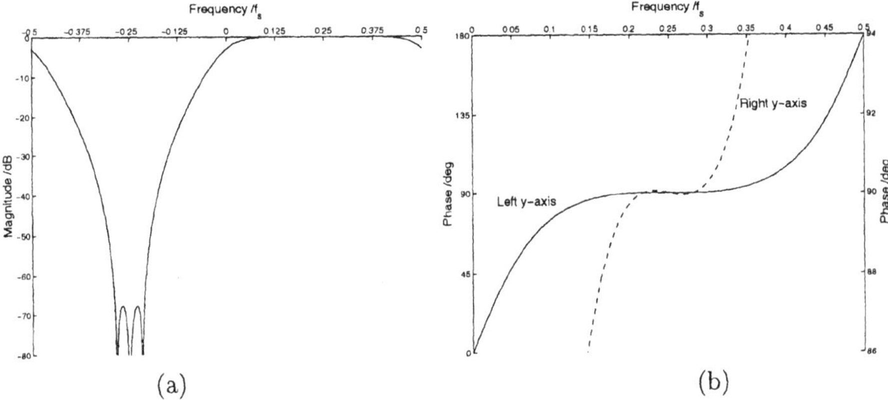

Figure 3.15: (a) Magnitude response ($|H_I(z)+jH_Q(z)|$) and (b) phase response ($\angle H_I(z) - \angle H_Q(z)$) in coarse and fine scale of the IIR Hilbert transformer of Example 3.7.

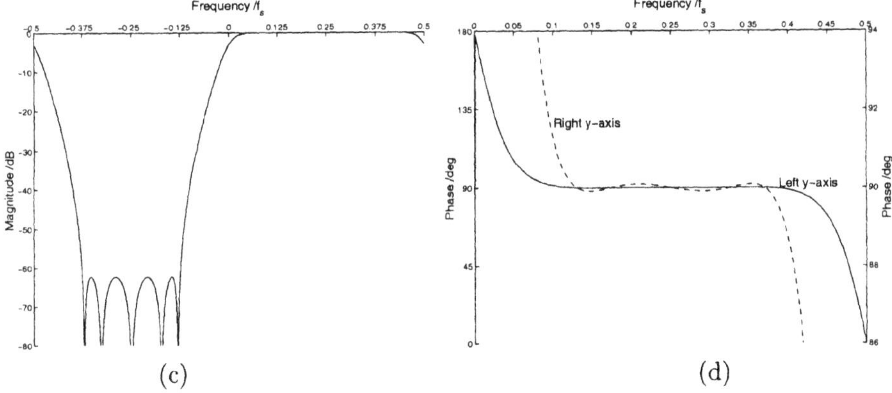

Figure 3.16: (a) Magnitude response ($|H_I(z)+jH_Q(z)|$) and (b) phase response ($\angle H_I(z) - \angle H_Q(z)$) in coarse and fine scale of the IIR Hilbert transformer of Example 3.8.

stop-band attenuation is 67 dB. Note that phase response is shown only on the positive frequency axis since it is odd symmetric.

□

Example 3.8. Consider the passband and stop-band width of 0.5π, and passband and stop-band ripple of less than 0.01 and 0.001 respectively. With the Matlab program, we get a transfer function with $N = 2$ and coefficients a_i, $1 \le i \le 2$, are 0.1380250 and 1.7103277. Therefore, the corresponding filter-I and filter-Q of the IIR Hilbert transformer are:

$$H_I(z) = z^{-1}\frac{z^{-2} - 0.5846832}{1 - 0.5846832z^{-2}}, \tag{3.47}$$

$$H_Q(z) = -\frac{z^{-2} - 0.1380250}{1 - 0.1380250z^{-2}}. \tag{3.48}$$

Its magnitude response ($|H_I(z) + jH_Q(z)|$) and phase response ($\angle H_I(z) - \angle H_Q(z)$) are shown in Figure 3.16. From the figure, we find that the actual stop-band attenuation is 62 dB. This transformer can be used in an efficient quadrature signal generator as described in [32].

□

Example 3.9. Consider the passband and stop-band width of 0.96π, and passband and stop-band ripple of less than 0.01 and 0.01 respectively. With the Matlab program, we get a transfer function with $N = 4$ and coefficients a_i, $1 \le i \le 4$, are 0.186542, 1.878465, 0.790201 and 1.061937. Therefore, the corresponding filter-I and filter-Q of the IIR Hilbert transformer are:

$$H_I(z) = z^{-1}\frac{z^{-2} - 0.532349}{1 - 0.532349z^{-2}}\frac{z^{-2} - 0.941675}{1 - 0.941675z^{-2}}, \tag{3.49}$$

$$H_Q(z) = -\frac{z^{-2} - 0.186542}{1 - 0.186542z^{-2}}\frac{z^{-2} - 0.790201}{1 - 0.790201z^{-2}}. \tag{3.50}$$

Its magnitude response ($|H_I(z) + jH_Q(z)|$) and phase response ($\angle H_I(z) - \angle H_Q(z)$) are shown in Figure 3.17. From the figure, we find that the actual stop-band attenuation is 40 dB. □

3.4 SC Hilbert Transformers

In this section, several switched-capacitor realizations of Hilbert transformers will be presented. The conventional two-phase circuit realization will be introduced first. Then three newly proposed polyphase circuits are discussed.

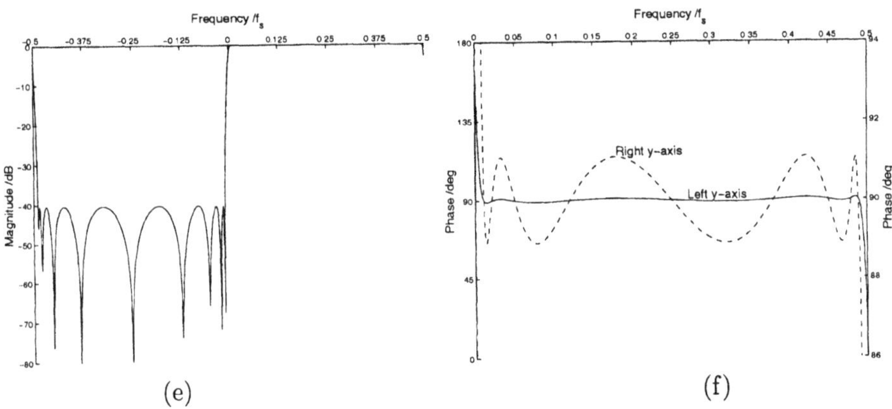

Figure 3.17: (a) Magnitude response ($|H_I(z)+jH_Q(z)|$) and (b) phase response ($\angle H_I(z) - \angle H_Q(z)$) in coarse and fine scale of the IIR Hilbert transformer of Example 3.9.

3.4.1 Two-phase Circuit

A two-phase SC circuit for IIR Hilbert transformer was proposed by Petraglia [33]. Figure 3.18 shows the circuit realizing the IIR Hilbert transformer illustrated in Example 3.8. The main building block of the transformer is an allpass section with transfer function $\frac{z^{-2}-a_i}{1-a_i z^{-2}}$. In Figure 3.18, the delay functions are realized by cascading half clock-period delay elements that employ unit gain buffer. And the filter coefficients a_i, $i = 1, 2$, are realized by the capacitance ratio C_1/C_2 and C_3/C_4. Note that the coefficients a_i in both numerator and denominator of $\frac{z^{-2}-a_i}{1-a_i z^{-2}}$ are realized by the same pair of capacitors. This makes its magnitude response insensitive to capacitor mismatches.

SWITCAP2 [34] circuit simulations have been conducted, and the resulted phase response under the assumption that capacitor ratio error of C_1/C_f is less that 0.1% is shown in Figure 3.19. We observe that this circuit approximates very well the ideal Hilbert transformer in a frequency band centered at $0.25f_s$ and with a bandwidth of $0.25f_s$ where the f_s is the clock frequency. The phase deviation from 90° is actually less than 0.15^o in that region.

The two-phase SC realization suffers from a number of problems. Firstly, unit-gain buffers are used therefore the transformer can not be realized in the differential form, which is an important technique to reject circuit noise. Secondly, the offset voltage in each delay elements will accumulate. This effect is referred as *offset error propagation.* Thirdly, the main opamps in the allpass

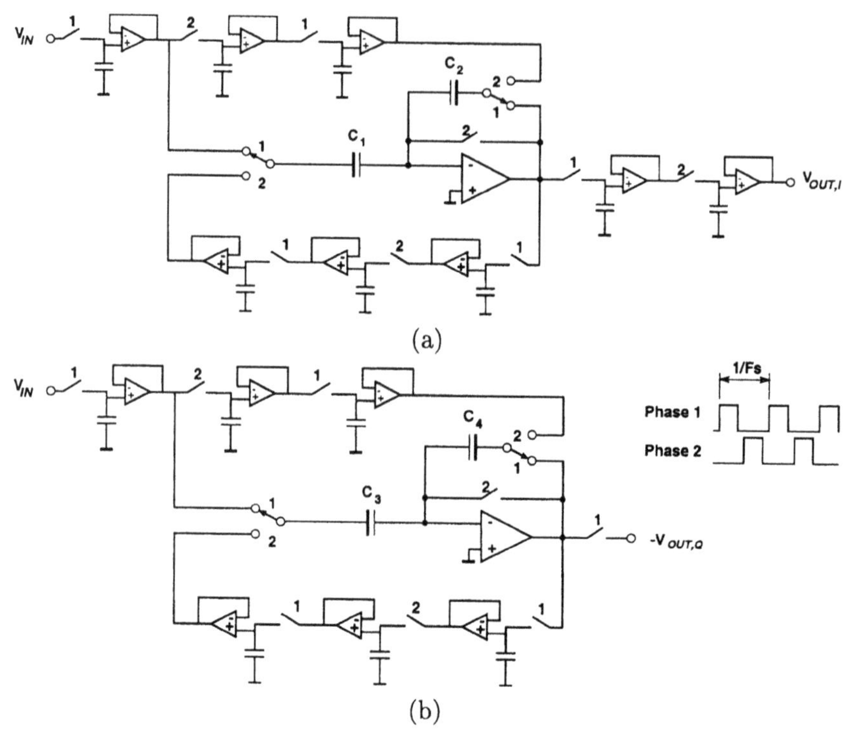

Figure 3.18: (a) Filter-I and (b) filter-Q of a two-phase SC IIR Hilbert transformer with the transfer function given in Example 3.8, where $C_1/C_2 = 0.5846832$, $C_3/C_4 = 0.1380250$ and capacitors without labelling are arbitrary holding capacitors.

sections are required to reset in every clock period. Thus fast opamps are demanded. Lastly, since many opamps/unit gain buffers are required, small chip area and low power consumption are difficult to achieve.

FIR circuit

The two-phase SC realization of FIR Hilbert transformer is straight- forward. As an example, the filter-I circuit of an FIR transformer with transfer function illustrated in example 3.4 ($H_I(z) = \alpha - z^{-2}$) is shown in Figure 3.20. The main difference from the IIR circuits is that there is no feedback path. Higher order FIR transformers can be obtained by connecting more SC branches to the inverting input terminal of the opamp.

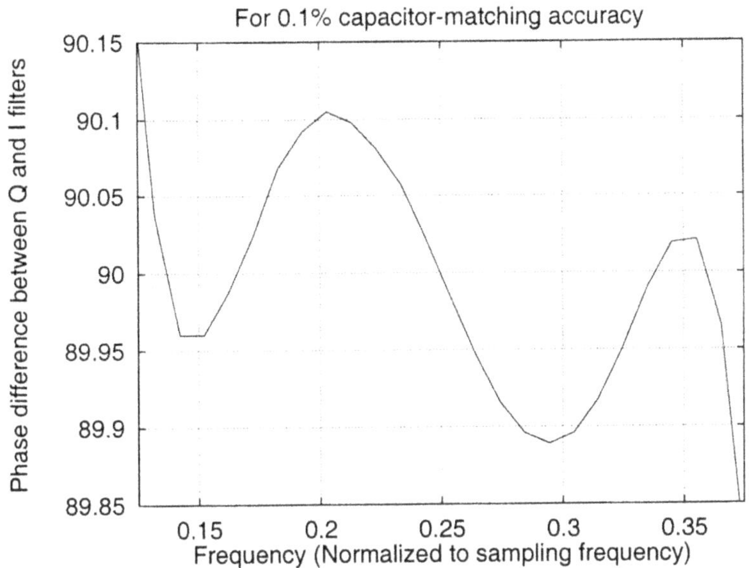

Figure 3.19: SWITCAP2 simulation results: Phase response of a two-phase SC IIR Hilbert transformer with the transfer function given in Example 3.8.

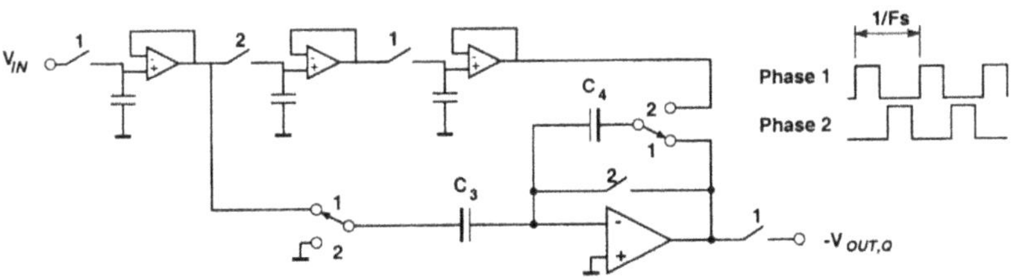

Figure 3.20: Circuit and timing diagram of the filter-I of a two-phase SC FIR Hilbert transformer, which has a transfer function of $H_I(z) = \alpha - z^{-2}$.

3.4.2 Polyphase Circuit

To solve the problems of the two-phase circuit, a polyphase SC implementation of IIR Hilbert transformers was proposed in [35]. Building blocks of the transformer, i.e, the unit delay z^{-1} and the allpass section $\frac{z^{-2}-a}{1-az^{-2}}$, realized by polyphase circuits are shown in Figure 3.21(a) and (c). The allpass section utilises a 1.5 clock-period polyphase delay circuit [36] as shown shown in Fig 3.21(b) to replace the cascaded unit buffer delay line used in the two-phase circuit.

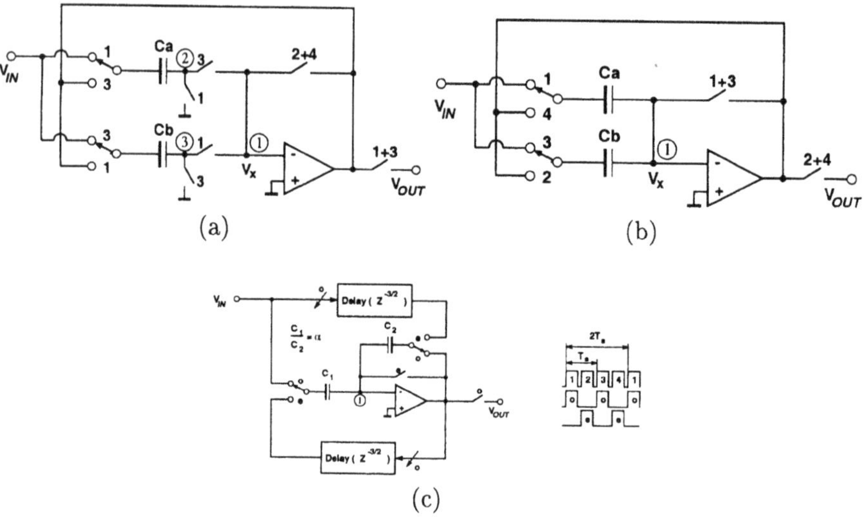

Figure 3.21: Polyphase building blocks of IIR Hilbert transformers: (a) unit-delay (z^{-1}); (b) 1.5-clock-period delay ($z^{-3/2}$) used in (c); (c) allpass section $\frac{z^{-2}-\alpha}{1-\alpha z^{-2}}$, where $\alpha = C_1/C_2$ and $T_s = 1/f_s$. Note that $C_a = C_b$ and are arbitrary hold capacitors.

The polyphase circuit inherits all the advantages of the two-phase circuit but avoids most of its problems. Firstly, it can be realized in differential form. Figure 3.22 shows the differential version of the allpass section. With differential circuit, common mode circuit noise such as clock feedthrough or charge injection error can be minimised. Secondly, there is no opamp offset error propagation. Because the cascaded delay lines are replaced by delay circuits employing only one opamp. Actually, the 1.5 clock-period delay circuit employed here is free of opamp offset voltage as explained in the following paragraph. Moreover, the delay circuits employed are also free of capacitor mismatches because the

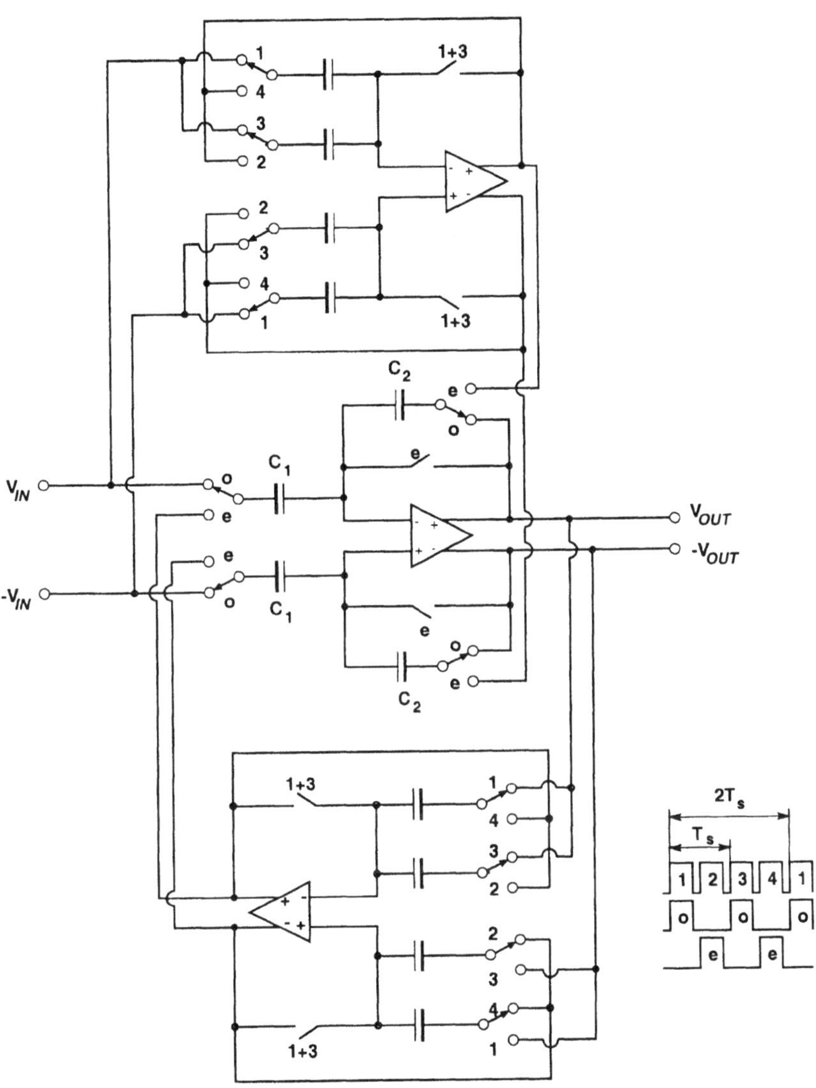

Figure 3.22: Fully differential polyphase SC allpass section $\frac{z^{-2}-a_i}{1-a_i z^{-2}}$, $a_i = C_1/C_2$, which constitutes the IIR Hilbert transformer.

input and output are sampled/produced by the same capacitor. Lastly, very few opamps are used. So considerable chip area and power dissipation can be saved.

The circuit shown in Figure 3.21(c) has immunity to opamp offset voltage. To explain this, let us model the offset voltage as an external source V_{off} in the positive terminal of an ideal opamp, and draw the circuit operating in different phases separately as shown in Figure 3.23(b) and (c). In clock phase e the output and inverting input of the opamp are short-circuited together and the capacitor C_1 and C_2 are charged to the input voltage V_2 and V_3 minus the offset voltage V_{off}. The charges stored in the top plate of C_1 and C_2 are

$$Q_1 = -C_1(V_2 - V_{off}), \quad \text{and} \tag{3.51}$$
$$Q_2 = -C_2(V_3 - V_{off}) \tag{3.52}$$

respectively. In phase o, the voltage at the inverting input terminal of opamp remains V_{off} due to the negative feedback through capacitor C_2. Capacitor C_1 is connected to input V_1 now. The charge transferred from top plate of C_1 to top plate of C_2 is

$$\begin{aligned} \Delta Q &= Q_1 - [-C_1(V_1 - V_{off})] \\ &= C_1(V_1 - V_2). \end{aligned} \tag{3.53}$$

So the charge accumulated in the top plate of C_2 is

$$\begin{aligned} Q_2' &= Q_2 + \Delta Q \\ &= -C_2(V_3 - V_{off}) + C_1(V_1 - V_2), \end{aligned} \tag{3.54}$$

and the output voltage is

$$\begin{aligned} V_{out} &= V_{off} - Q_2'/C_2 \\ &= V_{off} + (V_3 - V_{off}) - \frac{C_1}{C_2}(V_1 - V_2) \\ &= V_3 - \frac{C_1}{C_2}(V_1 - V_2). \end{aligned} \tag{3.55}$$

Therefore the output is not affect by the opamp offset. The delay circuit shown in Figure 3.21(b) is also insensitive to opamp offset. In clock phase 1 the output and inverting input of the opamp are short-circuited together and the hold capacitor C_a is charged to the input voltage minus the offset voltage V_{off}. During clock phase 4 the offset sample is subtracted from the instantaneous offset voltage of the opamp, so the output voltage will be exactly the input voltage delayed by 3 half clock period. However, the unit delay circuit shown in Figure 3.21(a) is affected by the opamp offset.

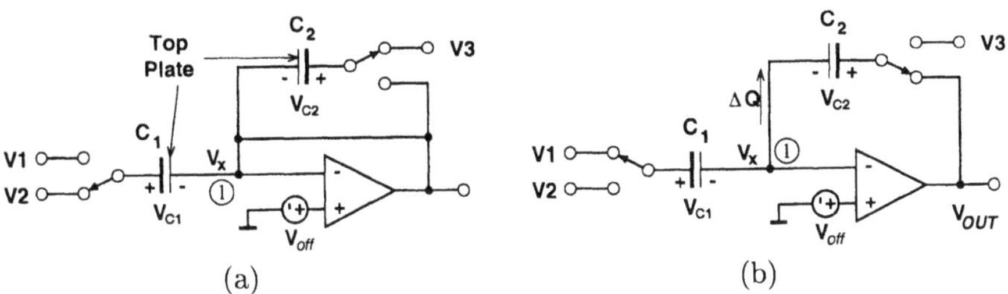

Figure 3.23: (a) Circuit of Figure 3.21(c) operating in phase e and (b) in phase o.

Finite opamp gain effect

The main limitation of the proposed circuit is that it is sensitive to finite opamp gain and bandwidth. To see how the finite gain affects the circuit performance, let us examine the effect on each building block firstly.

For a simple SC circuit, the transfer function can be written in the form [37, 38, 39]:

$$H(z) = H_i(z)g(z), \tag{3.56}$$

where $H_i(z)$ is the ideal (infinite opamp gain) transfer function and $g(z)$ is the error term due to the finite opamp gain.

Consider the unit delay shown in Figure 3.21(a), in the output phase o we have

$$\begin{aligned} V_x[nT] &= -\mu V_{out}[nT] && (3.57) \\ V_{out} &= V_x[nT] + V_{in}[(n-1)T] \\ &= -\mu V_{out}[nT] + V_{in}[(n-1)T] && (3.58) \\ \therefore V_{out}[nT] &= \frac{1}{1+\mu} V_{in}[(n-1)T] && (3.59) \\ \therefore H(z) &= z^{-1}\frac{1}{1+\mu}, && (3.60) \end{aligned}$$

where $\mu = 1/A$, A is the gain of the opamp, $T = 1/f_s$ is the sampling period and V_x is the voltage at the inverting input terminal of the opamp. From the above equation we have the error term g_1 of the unit delay circuit as $\frac{1}{1+\mu}$ which is independent of frequency. If parasitic capacitances are considered, it can be easily found that the error term $g_1(z)$ becomes

$$g_1(z) = \frac{1}{1+\mu(1+\frac{C_{p1}}{C})} \tag{3.61}$$

where C_{p1} represents the parasitic capacitance at node 1 and C represents the value of C_a and C_b for output at phase 4 and 2 respectively. It is seen that the presence of the parasitic capacitances slightly increases the magnitude of the error. Besides, the capacitor mismatch between C_a and C_b leads to slightly different value of $g_1(z)$ for phase 2 and 4, which is a higher order effect and can be actually neglected. This is the only way that the capacitance mismatches affect the performance of the circuit. In the rest of this chapter, this kind of mismatches will be ignored.

Following the same procedures, we can find that the error term of the delay circuit ($z^{-3/2}$) shown in Figure 3.21(b) has the same value of $\frac{1}{1+\mu}$. We use the same notation g_1 for the error term of this circuit. If parasitic capacitances are considered, it is same as that given in (3.61) except that C_{p1} should be replaced by $C_{p1} + C_{p2}$ or $C_{p1} + C_{p3}$ for output at phase 3 and 1 respectively, where C_{p2} and C_{p3} represent the parasitic capacitance at node 2 and 3. The mismatch between C_{p3} and C_{p2} is ignored, and we have the same expression of $g(z)$ for phase 1 and 3.

Now consider the core circuit of the allpass section. Let us refer to Figure 3.23 and ignore the opamp offset voltage to simplify the analysis (as explained before, it does not affect the output). During phase e, the inverting input and output of the opamp are short-circuited together. Denote the time index of this clock interval as $(n-0.5)T$, we have

$$\begin{aligned} V_x[(n-0.5)T] &= -\mu V_{out}[(n-0.5)T] \quad \text{and} \\ V_x[(n-0.5)T] &= V_{out}[(n-0.5)T], \end{aligned}$$

we have $V_x[(n-0.5)T] = 0$. Therefore $V_{c1}[(n-0.5)T] = V_2[(n-0.5)T]$ and $V_{c2}[(n-0.5)T] = V_3[(n-0.5)T]$. In the next phase, we have

$$\begin{aligned} V_x[nT] &= -\mu V_{out}[nT] \\ V_{c1}[nT] &= C_1(V_1[nT] - V_x[nT]). \end{aligned}$$

The inverting input of the opamp does not form a perfect virtual ground ($V_x \neq 0$). This causes an error in the charge $\Delta Q[nT]$ transferred to capacitor C_2:

$$\begin{aligned} \Delta Q[nT] &= V_{c1}[(n-0.5)T] - V_{c1}[nT] \\ &= C_1\{V_1[nT] - V_2[(n-0.5)T] - V_x[nT]\}. \end{aligned}$$

The capacitor voltage $V_{c2}[nT]$ and the output voltage $V_{out}[nT]$ are given by

$$\begin{aligned} V_{c2}[nT] &= V_{c2}[(n-0.5)T] - \Delta Q[nT]/C_2 \\ &= V_3[(n-0.5)T] - \frac{C_1}{C_2}\{V_1[nT] - V_2[(n-0.5)T] - V_x[nT]\}, \\ V_{out}[nT] &= V_x[nT] + V_{c2}[nT] \\ &= -\mu V_{out}[nT] + V_3[(n-0.5)T] - \frac{C_1}{C_2}\{V_1[nT] - V_2[(n-0.5)T] \\ &\quad - V_{out}[nT]\}, \\ V_{out}[nT] &= \frac{V_3[(n-0.5)T] - a(V_1[nT] - V_2[(n-0.5)T])}{1+\mu(1+a)}, \end{aligned} \tag{3.62}$$

where $a = \frac{C_1}{C_2}$. Therefore

$$V_{out}(z) = \frac{1}{1+\mu(1+a)}[-aV_1(z) + az^{-1/2}V_2(z) + z^{-1/2}V_3] \tag{3.63}$$

The above equation gives us the error term g_2 of the core of Figure 3.21(c) as $\frac{1}{1+\mu(1+a)}$. If parasitic capacitances are considered, the error term g_2 becomes:

$$g_2(z) = \frac{1}{1+\mu(1+\frac{C_1}{C_2}+\frac{C_{p1}}{C_2})}. \tag{3.64}$$

As a summary, the finite opamp gain error formulae of the circuits of Figure 3.21(a-c) are listed in Table 3.1.

In Figure 3.23, V_1 is the input, V_2 is the fed-back output with 1.5 clock-period delay, and V_3 is the input with 1.5 clock-period delay. Taking the finite gain effect of the delay circuit into account, we have

$$V_1(z) = V_{in}(z), \tag{3.65}$$

$$V_2(z) = g_1 z^{-3/2} V_{out}(z), \tag{3.66}$$

$$V_3(z) = g_1 z^{-3/2} V_{in}(z). \tag{3.67}$$

So, the output $V_{out}(z)$ in (3.63) becomes

$$V_{out}(z) = g_2[-aV_{in}(z) + az^{-2}g_1V_{out}(z) + z^{-2}g_1V_{in}(z)] \tag{3.68}$$

$$V_{out}(z) = \frac{z^{-2}(g_1g_2) - a(g_2)}{1 - az^{-2}(g_1g_2)}V_{in}(z). \tag{3.69}$$

And the transfer function of the allpass section $(\frac{z^{-2}-a}{1-az^{-2}})$ becomes

$$H(z) = \frac{z^{-2}(g_1g_2) - a(g_2)}{1 - az^{-2}(g_1g_2)}. \tag{3.70}$$

Table 3.1: Error formulae for compensated and uncompensated SC Hilbert Transformer Building Blocks with Finite Opamp Gain $A = 1/\mu$.

Circuits	Error term $g(z)$
Figure 3.21(a-b)	$1/(1+\mu)$
Figure 3.21(c)	$1/\left[1+\mu(1+\frac{C_1}{C_2})\right]$
Figure 3.27(a-b)	$1-\mu^2\left[1+\frac{C_h}{C_F}(1-z^{-1})\right]$
Figure 3.27(c)	$1+\mu^2(1+\frac{C_1}{C_2})\left[1+\frac{C_1'}{C_2'}+\frac{C_h}{C_2'}(1+z^{-1})\right]$
Figure 3.21(a-b)	$1/\left[1+\mu(1+\frac{C_{p1}}{C})\right]$
Figure 3.21(c)	$1/\left[1+\mu(1+\frac{C_1}{C_2}+\frac{C_{p1}}{C_2})\right]$
Figure 3.27(a-b)	$1-\mu^2\left(1+\frac{C_{p1}}{C}+\frac{C_{p2}}{C}+\frac{C_{p1}}{C_h}\right)\times \left(1+\frac{C_{p3}}{C_F}+\frac{C_{p1}+C_h}{C_F}(1-z^{-1})\right)$
Figure 3.27(c)	$1+\mu^2\left(1+\frac{C_1+C_{p1}+C_{p2}}{C_2}+\frac{C_{p1}}{C_h}+\frac{C_1C_{p1}}{C_2C_h}\right)\times \left(1+\frac{C_1'+C_{p3}}{C_2'}+\frac{C_h+C_{p1}}{C_2'}(1+z^{-1})\right)$

Note: The upper half of the table is the result without including the parasitic effect while the lower half of the table includes the effect.

Lastly, we get the finite opamp gain effect on the overall transfer function of an IIR Hilbert transformer, for example, the one illustrated in Example 3.8, as

$$\begin{cases} H_I(z) = \frac{z^{-2}(g_1g_{2i})-a_1(g_{2i})}{1-a_1z^{-2}(g_1g_{2i})}(g_1)z^{-1}, & a_1 = 0.5846832, \\ H_Q(z) = -\frac{z^{-2}(g_1g_{2q})-a_2(g_{2q})}{1-a_2z^{-2}(g_1g_{2q})}, & a_2 = 0.1380250. \end{cases} \tag{3.71}$$

where g_1 is the error term of the SC delay shown in Figure 3.21(a) and (c), g_{2i} and g_{2q} are the error terms of the allpass circuit with $a = 0.5846832$ and 0.1380250 respectively.

Computer simulations have been conducted to verify equation (3.71). The magnitude responses $|H_I + H_Q|$ of an SC transformer obtained by different methods are shown in Figure 3.24. The dotted-curves in the figure are obtained by evaluating equation (3.71). The circled-curves are SWITCAP2 simulation results. It is found that the two results match with each other exactly. In addition, two sets of results, one for opamp gain A=100 and the other for A=10000, are compared. It is observed that the response for A=10000 matches

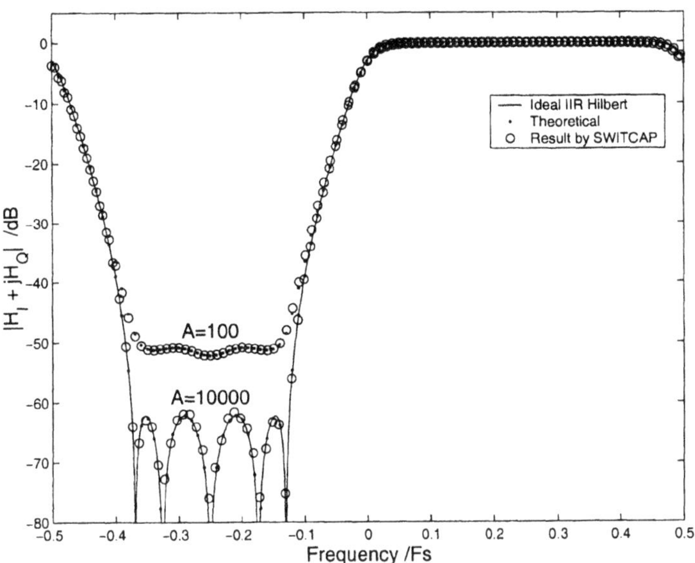

Figure 3.24: The magnitude response of the polyphase SC IIR Hilbert transformer illustrated in Example 3.8 with finite opamp gain A=100 and 10000. Dotted curves are obtained by evaluating equation (3.71) and circled curves are obtained by SWITCAP2 simulations.

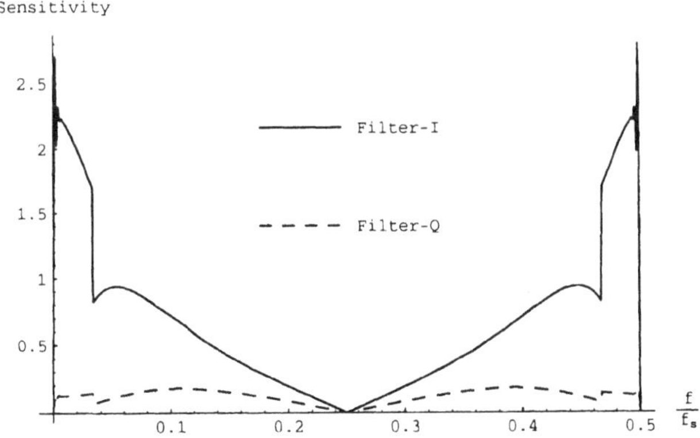

Figure 3.25: Phase sensitivities with respect to C_1/C_2 of filter-I (solid) and filter-Q (dashed) of the Hilbert transformer illustrated in Example 3.8.

very well with the ideal response. But the response for A=100 loses about 10 dB of attenuation in the stop-band.

Phase sensitivity to capacitor mismatch

It is also interesting to see how the circuit responds to the capacitance variation, i.e., the sensitivity to capacitance mismatches. Let us consider the magnitude response first. First, the unit delay circuit is insensitive to capacitor mismatches since its input and output signals are sampled or produced by the same capacitor. Second, the magnitude response of the allpass section $\frac{z^{-2}-a}{1-az^{-2}}$ is also insensitive to the capacitor mismatch. Since filter coefficient a in the numerator and denominator is realized by the same pair of capacitors, the magnitude response is independent of the value of a. However, the phase response of the allpass section does depend on the value of a, and thus the capacitance mismatch.

We define the phase sensitivity with respect to the parameter a as:

$$S_a^\theta = \frac{a}{|\angle H|}\frac{\partial|\angle H|}{\partial a}, \tag{3.72}$$

where $\angle H$ is the phase response. The phase sensitivities with respect to C_1/C_2 of the filter-I and -Q of the IIR Hilbert transformer are shown in Figure 3.25. From this figure, it is found that they are less than 0.5 and 0.175 in the care-band $[0.125f_s, 0.375f_s]$. Therefore, to obtain better than 0.1° phase accuracy, 0.2% and 0.57% capacitance matching accuracy of C_1/C_2 are required for filter-I and -Q respectively.

FIR circuit

The polyphase SC circuit realization of an FIR Hilbert transformer is straightforward. As an example, the filter-I circuit of an FIR transformer with transfer function illustrated in Example 3.4 ($H_I(z) = \alpha - z^{-2}$) is shown in Figure 3.26. The main difference from the IIR circuits is that there is no feedback path. Higher order FIR transformers can be obtained by connecting more SC branches to the inverting input terminal of the opamp.

3.4.3 Polyphase Circuit with Reduced Sensitivity to Opamp Gain and Bandwidth

The main limitation of the polyphase switched-capacitor Hilbert transformer described previously is the demand of high performance opamps. For high

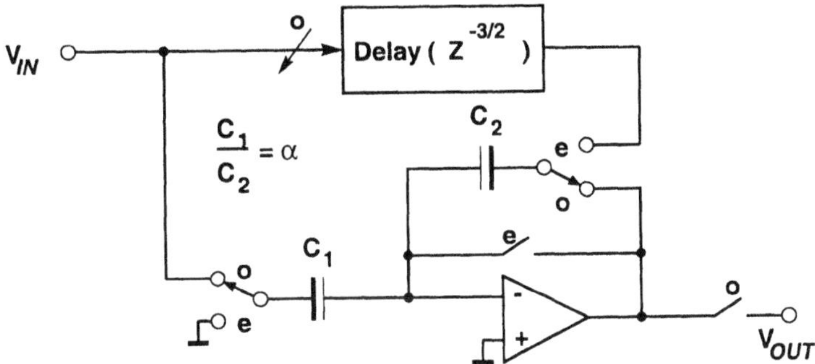

Figure 3.26: Circuit of the filter-I of a polyphase SC FIR Hilbert transformer, which has a transfer function of $H_I(z) = \alpha - z^{-2}$.

performance (over 60 dB stop band attenuation), the circuit requires high-gain ($> 70dB$) and high-bandwidth (unit-gain frequency 10 times the sampling frequency) opamps, which leads to large power consumption, especially for high frequency operation.

Traditional finite gain error compensation methods for SC filters exploit the auto-correlation property of the input signal [40, 41, 39, 42]. If the signal changes very little from sample to sample, then the finite gain error introduced during a clock interval can be compensated by making use of the knowledge about the finite gain error from the previous clock interval. But this principle does not apply to IIR Hilbert transformers that have signal frequency centered at one quarter of the sampling frequency, i.e., signal changes a lot from sample to sample.

Fortunately, a predictive correlated double sampling (CDS) technique [43] can be applied to the transformer to reduce its sensitivity to opamp gain and bandwidth, and therefore to extend its capability to high frequency applications [44].

In the predictive CDS circuit, a preliminary operation using a set of auxiliary capacitors which are matched to the main capacitors is performed during each clock interval. This preliminary operation provides us with a close approximation for the final gain error, which is stored and used for correction during the final operation. This scheme requires more chip area but achieves a very high precision.

The predictive CDS versions of SC building blocks of IIR Hilbert transformers are shown in Figure 3.27. In all these circuits, capacitor C_h is inserted to the inverting input of the opamp, and the reset switch of each opamp is replaced by a predictive SC branch. During clock phase e, the predictive CDS branch attempts to predict what the output voltage will be in the next clock interval (phase o). Therefore, the inverting terminal voltage of the opamp changes only negligibly from phase e to o. As a consequence, the finite gain effect which exhibits voltage variation in the inverting terminal will be cancelled. Note that the input of the circuit of Figure 3.27(c) is required to be held constant over phases e and o.

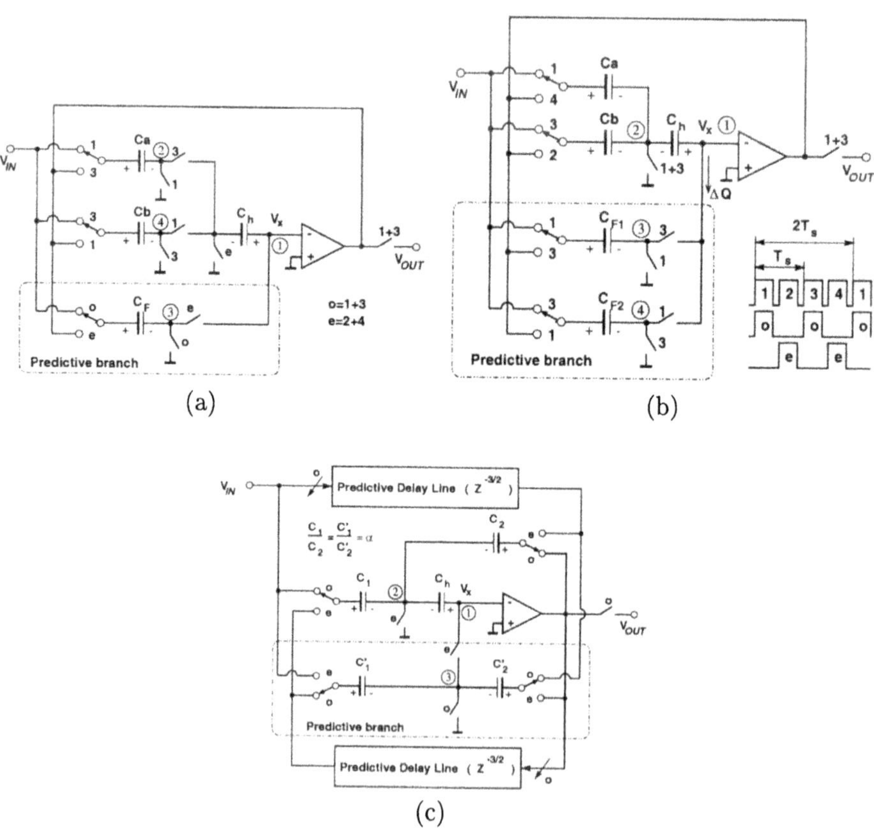

Figure 3.27: Gain and bandwidth compensated building blocks of polyphase SC IIR Hilbert transformers: (a)Predictive unit delay; (b)Predictive delay line ($z^{-3/2}$) used in (c); (c)Predictive all pass block $\frac{z^{-2}-\alpha}{1-\alpha z^{-2}}$, $\alpha = C_1/C_2$. Note that $C_a = C_b = C$, $C_{F1} = C_{F2} = C_F$ and they are arbitrary hold capacitors.

SC filters employing predictive CDS technique usually suffer from speed penalty because the fact that the input has to be held constant over two clock phases may necessitate the use of a third clock phase for coupling this circuit to other stages [43]. But in circuits of Figure 3.27 this penalty does not exist because they just utilise the otherwise unused opamp reset phase to generate the predictions.

Quantitative analysis of finite gain effect

To quantitatively analyse the finite gain effect on the predictive circuit of Hilbert transformers, let us consider the delay circuit shown in Figure 3.27(b) first. During phase 1, the input is sampled by both C_a and C_{F1} with respect to ground. Denote the time index of this clock interval as nT, we have

$$\begin{aligned} V_{Ca}[nT] &= V_{in}[nT] && (3.73) \\ V_{CF1}[nT] &= V_{in}[nT], && (3.74) \end{aligned}$$

where the polarity of the capacitor voltage is marked in the figure. During phase 2, the charges stored in C_a and C_{F1} are held constant since the capacitors are isolated. During phase 3 (time index $(n+1)T$), C_a is still holding its charges, and C_{F1} is connected to the inverting input and output of opamp to give a preliminary output voltage. This output suffers from an error due to the finite gain, giving rise to a nonzero voltage at inverting input V_x of opamp:

$$V_x[(n+1)T] = -\mu V_{out}[(n+1)T]. \tag{3.75}$$

While the top plate of capacitor C_h is always connected to the inverting input of opamp, the bottom plate is switched to ground during this phase, causing a charge transfer from C_h to C_{F1}:

$$\begin{aligned} \Delta Q &= C_h\{V_{Ch}[nT] - V_{Ch}[(n+1)T]\} \\ &= C_h\{V_x[(n-1)T] - V_x[(n+1)T]\}. && (3.76) \end{aligned}$$

The fact of $V_{Ch}[nT] = V_{Ch}[(n-1)T] = V_x[(n-1)T]$ used in the last step of above equation is because of that capacitor C_h holds its charge from phase 1 to 2 and from phase 3 to 4. The output voltage at time interval $(n+1)T$ (phase

3) is

$$\begin{aligned}
&V_{out}[(n+1)T] \\
&= V_x[(n+1)T] + V_{CF1}[nT] - \frac{C_h}{C_{F1}}\{V_{Ch}[(n)T] - V_{Ch}[(n+1)T]\} \\
&= -\mu V_{out}[(n+1)T] + V_{in}[nT] - \frac{C_h}{C_{F1}}\{V_x[(n-1)T] + \mu V_{out}[(n+1)T]\} \\
&\therefore V_{out}[(n+1)T] = \frac{1}{1+\mu+\mu k_h} V_{in}[nT] - k_h V_x[(n-1)T]. \qquad (3.77)
\end{aligned}$$

where $k_h = \frac{C_h}{C_{F1}}$. Let us denote $\beta = \frac{\mu}{1+\mu+\mu k_h}$ to simplify the analysis. From (3.75) and (3.77) we have

$$\begin{aligned}
V_x[(n+1)T] &= -\beta V_{in}[nT] + \beta k_h V_x[(n-1)T] \\
&= -\beta V_{in}[nT] + \beta k_h\{-\beta V_{in}[(n-1)T] + \beta k_h V_x[(n-2)T]\} \\
&= -\beta V_{in}[nT] - \beta^2 k_h V_{in}[(n-1)T] + \beta^2 k_h^2 V_x[(n-2)T]\} \\
&\cong -\beta V_{in}[nT] - \beta^2 k_h V_{in}[(n-1)T]. \qquad (3.78)
\end{aligned}$$

The last step above is obtained because $\beta^2 \ll 1$ and $V_x \ll V_{in}$. The voltage $V_x[(n+1)T]$ is stored in C_h.

During phase 4, the error voltage developed at V_x is very close to the value stored in C_h. A very good virtual ground is then produced at the bottom plate of C_h. Consequently, the error in the output voltage is significantly reduced. To analyse this quantitatively, we write

$$V_x[(n+1.5)T] = -\mu V_{out}[(n+1.5)T], \qquad (3.79)$$

$$\begin{aligned}
V_{out}[(n+1.5)T] &= V_x[(n+1.5)T] - V_{Ch}[(n+1)T] + V_{Ca}[nT] \\
&= V_x[(n+1.5)T] - V_x[(n+1)T] + V_{in}[nT] \\
&= -\mu V_{out}[(n+1.5)T] + (1+\beta)V_{in}[nT] \\
&\quad + \beta^2 k_h V_{in}[(n-1)T]. \qquad (3.80)
\end{aligned}$$

In $z-$domain, the above equation becomes

$$V_{out}(z) = -\mu V_{out}(z) + (1+\beta+\beta^2 k_h z^{-1})z^{-3/2}V_{in}(z). \qquad (3.81)$$

The result for output in phase 2 is exactly the same if $C_{F1} = C_{F2} = C_F$. Therefore, the transfer function of this circuit becomes:

$$\begin{aligned}
H(z) &= \frac{1+\beta+\beta^2 k_h z^{-1}}{1+\mu} z^{-3/2} \\
&= \left\{1 - \frac{\mu^2(1+k_h)}{(1+\mu)(1+\mu+k_h\mu)} + \frac{\mu^2 z^{-1}}{(1+\mu)(1+\mu+k_h\mu)^2}\right\} z^{-3/2} \quad (3.82)
\end{aligned}$$

where $z^{-3/2}$ is the ideal part and the first term is the error term $g(z)$ indicated in (3.56). Through some simplifications and ignoring the μ^3 and higher order terms, we have

$$g(z) \cong 1 - \mu^2 \frac{1 + \frac{C_h}{C_F}(1 - z^{-1})}{1 + \mu(2 + \frac{C_h}{C_F})} \cong 1 - \mu^2 \left[1 + \frac{C_h}{C_F}(1 - z^{-1})\right]. \quad (3.83)$$

Note that the error is proportional to μ^2. This makes it much smaller than the error in the circuit of Figure 3.21(b), which is proportional to μ. If parasitic capacitances are considered, the error term is given approximately by

$$g(z) = 1 - \mu^2 \left(1 + \frac{C_{p1}}{C} + \frac{C_{p2}}{C} + \frac{C_{p1}}{C_h}\right)\left(1 + \frac{C_{p3}}{C_F} + \frac{C_{p1} + C_h}{C_F}(1 - z^{-1})\right), \quad (3.84)$$

where C_{p1},C_{p2}, and C_{p3} represent the parasitic capacitances at nodes 1, 2, and 3 respectively, the parasitic capacitance at node 4 is assumed to be equal to C_{p3}, and C represents C_a and C_b. It is seen that the presence of the parasitic capacitances increases slightly the magnitude of the error. However, this circuit retains its relative advantage over the circuit of Figure 3.21(b) which is also slightly affected by the parasitic capacitances.

Following the same procedures, it can be found that the error term $g(z)$ of the unit delay circuit shown in Figure 3.27(a) is the same as that given in (3.83). If the parasitic capacitance is considered, the result is same as that given in (3.84) under the assumptions that $C_a = C_b = C$ and $C_{p4} = C_{p2}$.

The error term of the core of the allpass circuit shown in Figure 3.27(c) is given approximately by

$$g(z) = 1 + \mu^2(1 + \frac{C_1}{C_2})\left[1 + \frac{C_1'}{C_2'} + \frac{C_h}{C_2'}(1 + z^{-1})\right]. \quad (3.85)$$

If the parasitic capacitances are considered, the error term is given approximately by

$$g(z) = 1 + \mu^2 \left(1 + \frac{C_1 + C_{p1} + C_{p2}}{C_2} + \frac{C_{p1}}{C_h} + \frac{C_1 C_{p1}}{C_2 C_h}\right)\left(1 + \frac{C_1' + C_{p3}}{C_2'} + \frac{C_h + C_{p1}}{C_2'}(1 + z^{-1})\right).$$

These error formulae are summarised in Table 3.1. To see the overall impact of the finite gain on the Hilbert transformers, just substitute the corresponding value of $g(z)$ to equation (3.71).

Computer simulations have been conducted to verify the effectiveness of the predictive CDS SC Hilbert transformers. The magnitude responses $|H_I + H_Q|$ of the transformer obtained by different methods are shown in Figure 3.28.

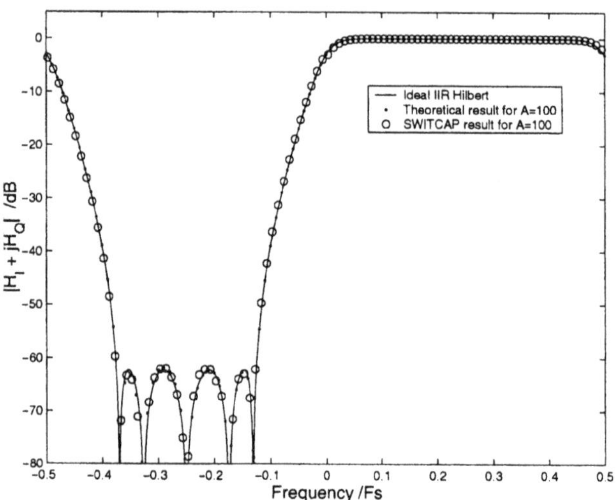

Figure 3.28: The magnitude response of the predictive CDS version of the polyphase SC IIR Hilbert transformer with finite opamp gain A=100 and $C_h/C_F = 1$. The dotted curve and circled curve are obtained by evaluating equation (3.71) and SWITCAP2 simulation respectively.

The dotted-curve in the figure is the result of evaluating equation (3.71). The circled-curve is SWITCAP2 simulation result. Once again, the two curves match with each other exactly. It can be clearly seen from the figure that a finite opamp gain of only 100 has virtually no effects on the circuit performance thanks to the use of predictive correlated-double-sampling technique.

Finite bandwidth effect

Finite opamp unit-gain bandwidth affects the settling time of switched-capacitor circuits. Although closed form solutions can be found for this effect on simple SC circuits, it is complicate to derive the closed form expression of this effect on the overall transfer function of a Hilbert transformer.

So we have conducted SWITCAP2 simulations to testify the finite bandwidth effect on the transformers. For opamp gain A=100 and unit-gain frequency $f_{GBW} = 3f_s$ (f_s is the clock frequency), Figure 3.29 shows that the gain and bandwidth compensated circuit has 8 dB more stop-band suppression than that of the uncompensated circuit described in the previous subsection.

As mentioned in the beginning of this subsection, the cost for this performance improvement is the use of more capacitors and switches, i.e., more chip area.

The reduced finite bandwidth effect of the predictive CDS Hilbert transformers can be understood in the following way. Remember that in the circuit introduced in the previous subsection, the opamp is reset during phase e. So the output must be charged from DC level to the final value during phase o. But in the predictive CDS circuit, two clock intervals are actually used for the output to be settled: during phase e the output is charged from the previous level to a preliminary level, then to the final value during phase o.

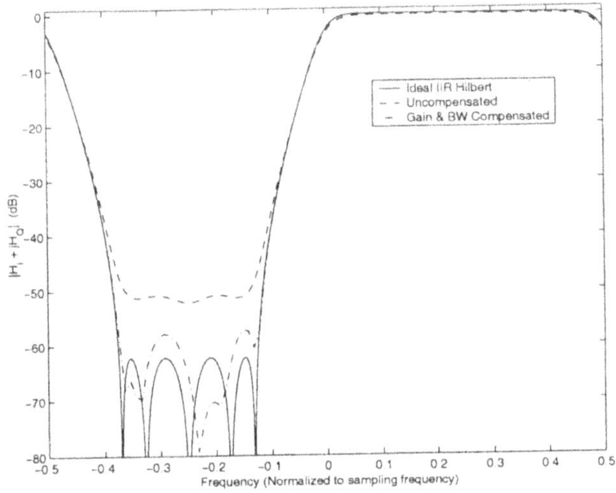

Figure 3.29: Comparison of the effect of finite gain and bandwidth on the frequency response of compensated and uncompensated SC Hilbert transformers, with opamp gain A=100 and unit-gain frequency $f_{GBW} = 3f_s$.

Other considerations

In the operation of circuits of Figure 3.27, the input offset voltage and other input-referred low-frequency noise components of the opamp are also stored in C_h during phase e and cancelled during phase o. Another important consideration is the clock-feedthrough. To minimise this effect, proper switching sequences can be employed and a fully differential configuration can be adopted. It is claimed in [43] that the predictive CDS technique itself offers a degree of cancellation of the clock feedthrough (both signal dependent and signal independent) by virtue of the predictive operation.

Lastly, the predictive CDS version of polyphase SC FIR Hilbert transformer is straight-forward to obtain and will not be demonstrated here.

3.4.4 Pseudo-N-Path Circuit

The concept of *pseudo-N-path* switched-capacitor circuits [45, 46, 47, 45] can be also used to realize discrete-time Hilbert transformers. The name of pseudo-N-path comes from the fact that only one physical path exists while the overall circuit represents various paths in different clock phases.

The delay circuits of Figure 3.21(a-b), in fact, belong to this class of circuits. Remember that in those circuits, outputs are produced by different capacitor branches in different clock phases. The allpass section $\frac{z^{-2}-\alpha}{1-\alpha z^{-2}}$ of the IIR Hilbert transformer can be also realized in pseudo-N-path form as shown in Figure 3.30(a) (only single-ended circuit is shown for simplicity). The circuit has 4 parallelled switched-capacitor branches, each one samples the input/output signals in a specific phase, holds the values in next three phases and gives the output four phases later in the form of $V_{out}[nT] = V_{in}[(n-2)T] - \alpha V_{in}[nT] - \alpha V_{out}[(n-2)T]$, where $T = 1/f_s$ is the sampling period. A total of 8 phases at lower frequency ($\frac{1}{4}f_s$) are used and there is no extra speed requirements on the opamp.

The biggest advantage of circuit of Figure 3.30(a) over the previous circuit is that it employs only one opamp. Thus power consumption can be greatly reduced, especially for high frequency applications. The main drawback is its sensitivity to capacitor mismatches between different SC branches.

The finite opamp gain effect on the circuit of Figure 3.30(a) is same as that of the circuit shown in Figure 3.21(c), which is given by (3.64). The predictive correlated-double-sampling technique can also be used in this circuit to reduce the finite gain and bandwidth effects. It is straight-forward to find the predictive CDS version of the circuit of Figure 3.30 and we will not discuss it here.

SWITCAP2 simulations were conducted for a pseudo-N-path SC IIR Hilbert transformer with transfer function $H_I(z) = z^{-1}$, $H_Q(z) = \frac{1-3z^{-2}}{3-z^{-2}}$. Note that the coefficients of the transformer are all integers. The magnitude response is shown in Figure 3.31 for the opamp gain $A = 10000$. The simulated result coincides with the ideal one, which verifies the correctness of the circuit.

Pseudo-N-path SC circuit is also capable of realizing an FIR Hilbert transformer. As an example, Figure 3.30(b) shows the circuit of filter-I of an FIR transformer with transfer function illustrated in Example 3.4 ($H_I(z) = \alpha - z^{-2}$).

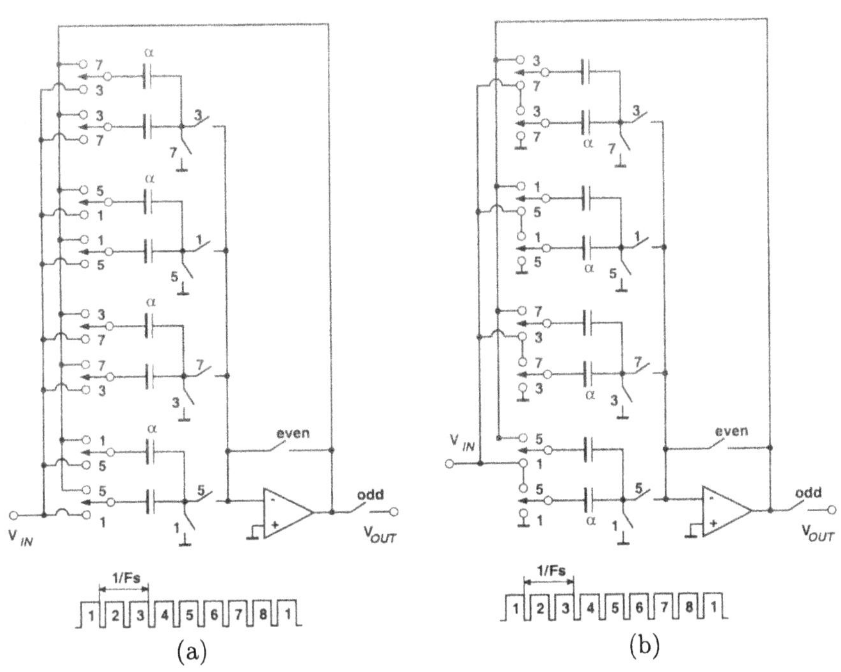

Figure 3.30: Circuit and timing diagram of pseudo-N-path SC circuits of (a) allpass section $\frac{z^{-2}-\alpha}{1-\alpha z^{-2}}$ for IIR Hilbert transformers; (b) the filter-I of an FIR Hilbert transformer, which has a transfer function of $H_I(z) = \alpha - z^{-2}$.

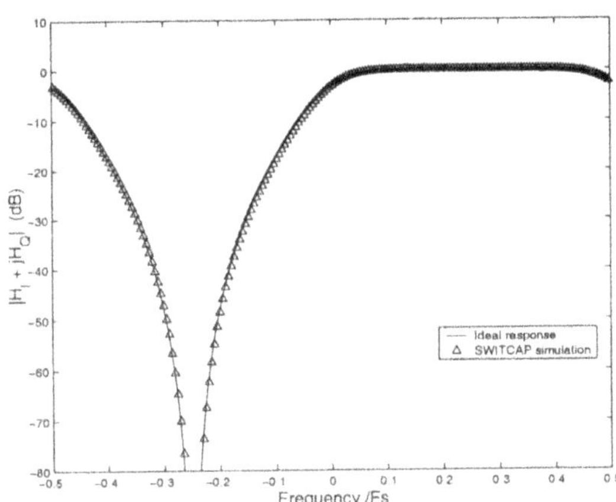

Figure 3.31: Magnitude of a pseudo-N-path SC IIR Hilbert transformer with opamp gain=10000.

FIR circuits with higher order of z^{-1} can be easily obtained by parallelling more SC branches.

3.5 Summary

In this chapter we have presented design methods of Hilbert transformers, i.e., 90° phase shifters, and their circuit realizations in both continuous-time and discrete-time domains.

The traditional way to realize an analog Hilbert transformer is to use continuous-time circuit techniques such as passive and active RC networks as well as polyphase RC network. A main issue of these CT circuit realizations is that tuning is usually required because the edge frequencies of the transformers depend on the absolute values of R and C. Otherwise the care-band bandwidth must be over-designed.

The more accurate way to realize an Hilbert transformer is to employ discrete-time circuit techniques, mainly the switched-capacitor technique. Several SC realizations have been presented. The first one is a conventional two-phase SC circuit, which employs many unit gain buffers and suffers from a number of problems. The second one is a newly proposed polyphase SC circuit, which inherits all the advantages of the previous circuit but avoids most of its drawbacks. This circuit is free of offset error and capacitor mismatches (in the delay circuits), and has fewer opamps. The remaining problem is the high sensitivity to opamp finite gain. To solve it, the predictive correlated-double-sampling technique is used. The new circuit reduces significantly the sensitivity to finite opamp gain as well as finite opamp bandwidth. The last alternative is a pseudo-N-path circuit, which employs fewest number of opamps, thus consumes less power. However, it suffers from the problem of capacitance mismatches.

These Hilbert transformers can be used in many wireless communication systems where wideband 90° phase shifting is required.

References

[1] K. Yamamoto, K. Maemura, N. Andoh, and Y.Mitsui, "A 1.9GHz band GaAs direct-quadrature modulator IC with a phase shifter," *IEEE J. Solid-State Circuits*, vol. 28, no. 10, pp. 994–1000, Oct 1993.

[2] A. Boveda, F. Ortigoso, and J.I. Alonso, "A 0.7-3GHz GaAs QPSK/QAM direct modulator," in *Digest of Technical Papers, IEEE Int. Solid-State Circuit Conference*, Feb. 1993, pp. 142–143.

[3] S. Otaka, T. Yamaji, C.Takahashi, and H. Tanimoto, "A low local input 1.9GHz Si-bipolar quadrature modulator with no adjustment," *IEEE J. Solid-Sate circuits*, vol. 31, no. 1, pp. 30–37, Jan 1996.

[4] T. Tsukahara, M. Ishikawa, and M. Muraguchi, "A 2V 2GHz Si-bipolar direct-conversion quadrature modulator," *IEEE J. Solid-State Circuits*, vol. 31, no. 2, pp. 262–267, Feb 1996.

[5] M. Steyaert and R. Roovers, "A 1-GHz single-chip quadrature modulator," *IEEE J. Solid-State Ciruits*, vol. 27, no. 8, pp. 1194–1197, Aug 1992.

[6] T.D. Stetzler, I.G. Post, J.H. Havens, and M. Koyama, "A 2.7-4.5V single chip GSM transceiver RF integrated circuit," *IEEE J. Solid-State Circuits*, vol. 30, no. 12, pp. 1421–1429, Dec 1995.

[7] Stefan L. Hahn, *Hilbert Transforms in Signal Processing*, Artech House, 1996.

[8] Alan V. Oppenheim and Ronald W. Schafer, *Discrete-time Signal Processing*, Prentice Hall, 1989.

[9] I.A. Koullias, J.H. Havens, I.G. Post, and P.E. Bronner, "A 900MHz transceiver chip set for dual-mode cellular radio mobile terminals," in *Digest of Technical Papers, IEEE Int. Solid-State Circuit Conference*, Feb. 1993, pp. 140–141.

[10] M. H. Ackroyd, "Simplified quadrature network," *Electronic Letters*, vol. 4, pp. 105–107, 1968.

[11] S.D. Bedrosian, "Normalised design of 90 phase-difference networks," *Trans. of the Inst. Radio Engrs*, vol. CT-7, pp. 128–136, 1960.

[12] S. Darlington, "Realisation of a constant phase difference," *Bell Syst. Tech. J.*, vol. 29, pp. 94–104, 1950.

[13] R. B. Dome, "Wide-band phase shift networks," *Electronics*, vol. 19, pp. 112–115, Dec. 1946.

[14] G. G. Gourier and G. F. Newell, "A quadrature network for generating vestigial sideband signals," *Proc. Inst. Electr. Engrs*, vol. 107B, pp. 253–260, 1960.

[15] H. J. Orchard, "Synthesis of wide-band two-phase networks," *Wireless Engr.*, vol. 27, pp. 72–81, 1950.

[16] H. J. Orchard, "The synthesis of RC networks to have prescribed transfer functions," *Proc. IRE*, vol. 39, pp. 428–432, Apr. 1951.

[17] W. Saraga, "The design of wideband two-phase networks," *Proc. Inst. Electron. Engrs*, vol. 38, pp. 754–770, 1950.

[18] D. K. Weaver, "Design of RC wide-band 90-degree phase difference networks," *Proc. IRE*, vol. 42, pp. 671–676, Apr. 1954.

[19] M. Faulkner, T. Mattsson, and W. Yates, "Automatic adjustment of quadrature modulators," *IEE Electronic Letters*, vol. 27, no. 3, pp. 214–216, Jan. 1991.

[20] M. D. McDonald, "A 2.5GHz BiCMOS image-reject front end," in *Digest of Technical Papers, IEEE Int. Solid-State Circuit Conference*, 1993, pp. 144–145.

[21] D. Pache, J.M. Fournier, G. Billiot, and P. Senn, "An improved 3V 2GHz BiCMOS image reject mixer IC," in *Proc. IEEE Custom Integrated Circuits Conference*, 1995, pp. 95–98.

[22] J. Crols and M. Steyaert, "A fully integrated 900 MHz CMOS double quadrature downconverter," in *Digest of Technical Papers, IEEE Int. Solid-State Circuit Conference*, 1995, pp. 136–137.

[23] F. J. Kub and E. W. Justh, "Analog CMOS implementation of high frequency least-mean square error learning circuit," *IEEE J. Solid-State Circuits*, vol. 30, no. 2, pp. 1391–1398, Dec. 1995.

[24] D. J. Comer and J. E. McDermid, "Inductorless bandpass characteristics using allpass networks," *IEEE Trans. Circuit Theory*, vol. CT-15, pp. 501–503, Dec. 1968.

[25] G. S. Moschytz, "A high Q, insensitive active RC network, similar to the Tarmy-Ghausi circuit, but using single-ended operatoinal amplifiers," *Electron. Lett.*, vol. 8, pp. 458–459, Sept. 1972.

[26] D.T. Comer, "A wideband, fixed gain BiMOS amplifier," in *Proc. Int. ASIC Conf. Rochester*, Rochester, NY, USA, Aug. 1993, pp. 27–30.

[27] L. B. Jackson, "On the relationship between digital Hilbert transformers and certain low-pass filters," *IEEE Trans.*, vol. ASSP-24, pp. 381, 1976.

[28] L. B. Jackson, "On the relationship between digital Hilbert transformers and certain low-pass filters," *IEEE Trans.*, vol. ASSP-33, pp. 381–383, Aug. 1985.

[29] J.E. Eklund and R. Arvidsson, "A multiple sampling, single A/D conversion technique for I/Q demodulation in CMOS," *IEEE J. Solid-State Circuits*, vol. 31, no. 12, pp. 1987–1994, Dec. 1996.

[30] Rashid Ansari, "IIR discrete-time hilbert transformers," *IEEE Trans. on Acoustics, Speech and Signal Processing*, vol. ASSP-35, no. 8, pp. 1116–1119, Aug. 1987.

[31] Rashid Ansari, "Elliptic filter design for a class of generalized halfband filters," *IEEE Trans. on Acoustics, Speech and Signal Processing*, vol. ASSP-33, no. 4, pp. 1146–1150, Oct. 1985.

[32] Charles M. Rader, "A simple method for sampling in-phase and quadrature components," *IEEE Trans. on Aerospace and Electronic Systems*, vol. AES-20, no. 6, pp. 821–824, Nov. 1984.

[33] A. Petraglia, F.A.P. Baruqui, and S.K. Mitra, "Recursive switched-capacitor Hilbert transformers," in *Proc. Int. Symposium on Circuits and Systems*, 1998, vol. 1, pp. 496–499.

[34] K. Suyama and S.C. Fang, *User's Manual for SWITCAP2 version 1.1*, Columbia University, 1992.

[35] K.P. Pun, J.E. Franca, and C. Azeredo Leme, "Polyphase SC IIR Hilbert transformers," *IEE Electronics Letters*, vol. 35, no. 9, pp. 689–670, 29th April 1999.

[36] K. Matsui, T. Matsuura, S. Fukasawa, Y. Izawa, Y. Toba, N.Miyake, and K. Nagasawa, "Cmos video filters using switched capacitor 14mhz circuits," *IEEE J. Solid-State Circuits*, vol. SC-20, pp. 1096–1102, Dec. 1985.

[37] G. C. Temes, "Finite amplifier gain and bandwidth effects in switched-capacitor filters," *IEEE J. Solid-State Circuits*, vol. SC-15, pp. 358–361, June 1980.

[38] K. Martin and A. S. Sedra, "Effects of finite gain and bandwidth on the performance of switched-capacitor filters," *IEEE Trans. Circuits and Systems*, vol. CAS-28, pp. 822–829, 1981.

[39] K. Haug, F. Maloberti, and G. C. Temes, "Switched-capacitor integrators with low finite-gain sensitivity," *IEE Electronic Letters*, vol. 21, no. 24, pp. 1156–1157, 1985.

[40] G. Fischer and G. S. Moschytz, "SC filters for high frequencies with compensation for finite gain effect," *IEEE Trans. Circuits and Systems*, vol. CAS-32, pp. 1050–1056, Oct. 1985.

[41] K. Nagaraj, K. Singhal, T. R. Viswanathan, and J. Vlach, "Reduction of finite-gain effect in switched-capacitor filters," *IEE Electronic Letters*, vol. 21, pp. 644–645, 1985.

[42] K. Haug, F. Maloberti, and G. C. Temes, "Switched-capacitor circuits with low opamp gain sensitivity," in *Proc. IEEE Int. Symposium on Circuits and Systems*, San Jose, CA, USA, May 1985, pp. 797–800.

[43] K. Nagaraj, T.R. Viswanathan, K.Singhal, and J.Vlach, "Switched-capacitor circuits with reduced sensitivity to amplifier gain," *IEEE Trans. Circuits and Systems*, vol. CAS-34, no. 5, May 1987.

[44] K.P. Pun, J.E. Franca, and C. Azeredo Leme, "Polyphase SC IIR Hilbert transformer with reduced sensitivity to finite gain and bandwidth," *IEE Electronics Letters*, vol. 35, no. 19, pp. 1602–1603, 16th September 1999.

[45] Teng-Hsien Hsu and G. C. Temes, "An improved circuit for pseudo-N-path switched-capacitor filters," *IEEE Trans. Circuits and Systems*, vol. CAS-32, pp. 1071–73, Oct. 1985.

[46] J. Pandel et al., "Integrated 18th-order pseudo-N-path filter in VIS-SC technique," *IEEE J. Solid-State Circuits*, vol. SC-21, pp. 48–55, Feb 1986.

[47] Jung-Chen Lin and J. H. Nevi, "Differential charge-domain bilinear-Z switched-capacitor pseudo-N-path filters," *IEEE Trans. Circuits and Systems*, vol. CAS-35, pp. 409–414, April 1988.

Appendix 3.A

```
% HB2HILB  IIR Hilbert filter design through half-band filter
%   [A,As]=HB2HILB(Wp,Rp,Rs) return filter coeficients A and
%   actual stop-band attenuation As (dB).
%   Input Wp is pass-band edge, in unit of pi;
%   Rp and Rs are riples of pass-band and stop-band.
%   Example: hb2hilb(0.075,0.1,0.001)
```

```
function [a,As]=hb2hilb(Wp,Rp,Rs)
  Wp=Wp*pi; Ws=(pi-Wp);
  delta=min(Rs,sqrt(2*Rp-Rp^2));
  Delta=(1/delta^2-1);
  k=(tan(Wp/2))^2;
  k2=sqrt(1-k^2);
  p=(1-sqrt(k2))/(1+sqrt(k2))/2;
  q=p+2*p^5+15*p^9+150*p^13;
  tmp=ceil(-2*log(4*Delta)/log(q));
  if 2*floor(tmp/2)==tmp,
    N=tmp-1;
  else
    N=tmp;
  end
  L=(N-1)/2;
  if L<1,
    error('Wrong Wp, Rp or Rs value');
  end
  for i=1:L,
    tmp1=sin(pi*i/N);            tmp2=0;
    tmp3=sin(pi*(2*i-1)/2/N);    tmp4=0;
    for m=1:100,
        tmp1=tmp1+(-1)^m*q^(m*(m+1))*sin((2*m+1)*pi*i/N);
        tmp2=tmp2+(-1)^m*q^(m*m)*cos(2*m*pi*i/N);
        % following is for determining peak frequency
        tmp3=tmp3+(-1)^m*q^(m*(m+1))*sin((2*m+1)*pi*(2*i-1)/2/N);
        tmp4=tmp4+(-1)^m*q^(m*m)*cos(m*pi*(2*i-1)/N);
    end
    Omega(i)=(2*q^0.25)*tmp1/(1+2*tmp2);
    w_peak(i)=(2*q^0.25)*tmp3/(1+2*tmp4);
  end
  r=sqrt((1-Omega.^2*k).*(1-Omega.^2/k));
  for i=1:L,
    c_phi(i)=(-1)^(i+1)*r(i)/(1+Omega(i)^2);
  end
  a=(1-c_phi)./(1+c_phi);
  epsilon=(prod(w_peak))^2*sqrt(k);
  As=20*log10(1/sqrt(1+1/epsilon^2));
return;
```

Chapter 4

Sampled-Data Image-Rejection Receiver

4.1 Introduction

In this chapter, we will address the applications of switched-capacitor (SC) Hilbert transformers which are presented in the previous chapter. The main function of SC Hilbert transformers is wideband 90° phase shifting. Remember that a 90° phase shifter is required in Hartley image rejection receiver for performing on-chip image rejection. Ideally, complete image cancellation can be achieved in a Hartley receiver. However, due to mismatches between I and Q paths and phase-shifting error in the 90° phase shifter, the receiver offers limited image rejection performance. Only up to about 35 dB of image rejection ratio (IRR) has been reported [1, 2, 3, 4].

The image rejection performance of the Hartley receiver can be improved by improving the accuracy of the 90° phase shifter, which which must perform the phase shifting over the whole signal bandwidth. The phase shifter is implemented by passive or active RC networks conventionally, but it is difficult to achieve high gain and phase accuracy in a broad frequency band. Instead, it can be implemented by SC Hilbert transformers with very high accuracy [5]. This novel form of the realization of the Hartley receiver is called as the sampled-data image rejection receiver (SDIRRx).

The architecture and applications of the SDIRRx is presented in this chapter. Moreover, the design of an SDIRRx in 0.6μm CMOS technology and some experimental results are presented.

4.2 Architecture of the Sampled-Data Image Rejection Receiver

Figure 4.1(a) shows the block diagram of a Hartley receiver while the block diagram of the proposed sampled-data image rejection receiver is shown in Figure 4.1(b). The SDIRRx contains two matched mixers, two matched anti-aliasing filters and a pair of SC Hilbert transformer performing 90° phase shift. The summing circuit is embedded in the SC circuit. The sampling frequency f_s of the transformer is four times the output center frequency since the transformer has its "care-band" around $f_s/4$.

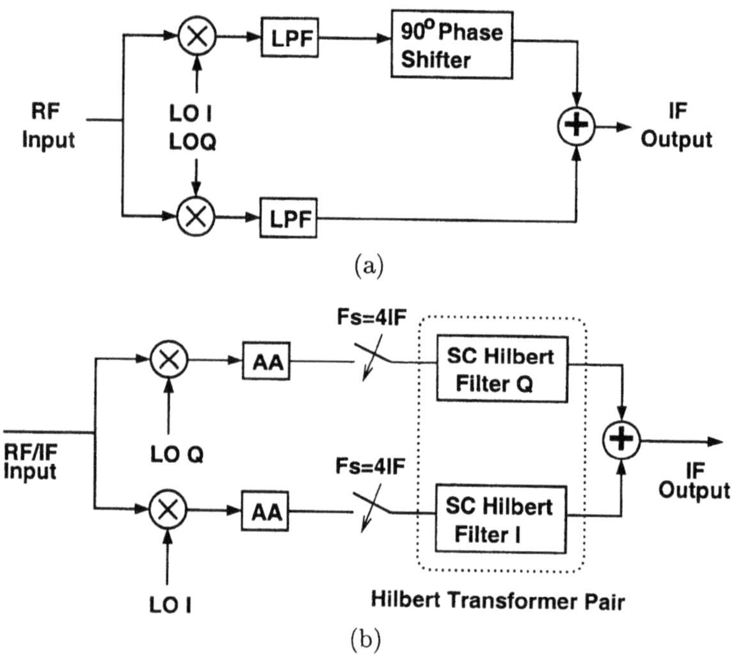

Figure 4.1: Architectures of (a) Hartley image rejection receiver and (b) the sampled-data image rejection receiver.

The SDIRRx produces discrete-time IF output, which can be directly converted to digital signal by an IF sampling A/D converter for further processing. The IF A/D conversion avoids DC offset and 1/f noise which is encountered in baseband sampling receivers. Yet, with an image rejection section in front of the IF sampling reduces the anti-aliasing filtering requirement, and allows a high IF (input) for easy front-end design [9]. In [9], the simple first order SC Hilbert transformer is used and thus the ratio of sampling frequency to signal bandwidth must be very high. If a higher order SC Hilbert transformer is used, the ratio can be greatly reduced.

On the other hand, the discrete-time IF output of the SDIRRx can also be converted back to continuous-time domain for further processing. In the next section, such an application will be discussed in details.

The IRR of the Hartley receiver is governed by:

$$IRR = \frac{1 + 2e_g \cos\varepsilon + e_g^2}{1 - 2e_g \cos\varepsilon + e_g^2} \tag{4.1}$$

where e_g is the gain ratio between I and Q paths, and ε represents the phase imbalance between the two paths. The gain and phase imbalances are contributed by:

- mismatch between I and Q mixers;
- gain and phase mismatch between the I and Q LO;
- mismatch between two lowpass filters;
- gain and phase error of the phase shifter in the signal path.

In other words, every circuit component in the two paths contributes gain and phase errors.

The mismatches between I and Q mixers and between two lowpass filters arise from local electrical characteristic variations and geometry alignment errors generated the fabrication process. Careful layout is essential to achieve high matching accuracy.

The gain and phase errors of the quadrature LO signal are relatively easy to control. Quadrature LO signal can be generated by the (1) use of oscillators with inherent quadrature output; (2) use of a frequency divider with one output triggered by the rising edge and another by the falling edge; (3) use of a phase shifter. The first method is the best approach. A phase error of 0.5^o and magnitude imbalance of 1% [6](corresponding to 45 dB IRR) can be easily achieved. The performance of the second method is affected by the duty cycle of the clock signal [1]. The third method suits for applications with narrow LO tuning range [7, 8].

The dominant error source is the 90° phase shifter in the signal path. This phase shifter is traditionally realized by passive or active RC/CR networks. Since the absolute values of resistance and capacitance in an integrated circuit are sensitive to temperature and have large deviation from the design value, it is difficult to control the phase and magnitude balance over the whole signal band. In the SDIRRx, more accurate and better controlled SC 90° phase shifter is used.

4.3 Design of an SDIRRx in 0.6μm CMOS

An SDIRRx was designed for cordless telephone applications, according to CT-1 standard [10]. CT-1 is an analog system adopting FM modulation/demodulation technique. Most of the CT-1 chips available in the current market employ several tens of external passive components including image rejection filters [11]. The SDIRRx was designed with the goals to eliminate an expensive off-chip image rejection filter while keeping high image rejection performance, and to be compatible with most CT-1 standards around the world. The design of the SDIRRx in a 0.6μm CMOS technology is presented in detail in the following subsections.

4.3.1 System Architecture

Figure 4.2 shows the system architecture of the receiver, where the shadowed blocks were integrated in a 0.6μm CMOS technology. Figure 4.3 illustrates the frequency spectra developed at RF, IF_1 and IF_2. Table 4.1 gives the frequency allocation.

Table 4.1: Frequencies at different points of the receiver

RF	20-50 MHz
LO_1	10 MHz-40 MHz (rough tuning)
IF_1	centered at 10.432 MHz, bandwidth = 500 kHz
LO_2	10 MHz-10.48 MHz (fine tuning)
IF_2	192 kHz (signal bandwidth=20 kHz)

The RF signal received from the antenna is first filtered by an RF preselection filter. Since the first IF is high enough (above 10MHz), the requirement of this filter is not stringent. The next step is low-noise amplification, and then the mixing process to down-convert the signal to the first IF. Note that the carrier frequencies of different CT-1 standards range from 20 MHz to 50 MHz (except U.K. base-set). But in each country, the range is not wider than 25 times of 20 kHz signal bandwidth, i.e, 500 kHz. After low noise amplification, the whole frequency band of 25 consecutive channels in a specific country is mixed down to the first IF centered at 10.432 MHz by the first

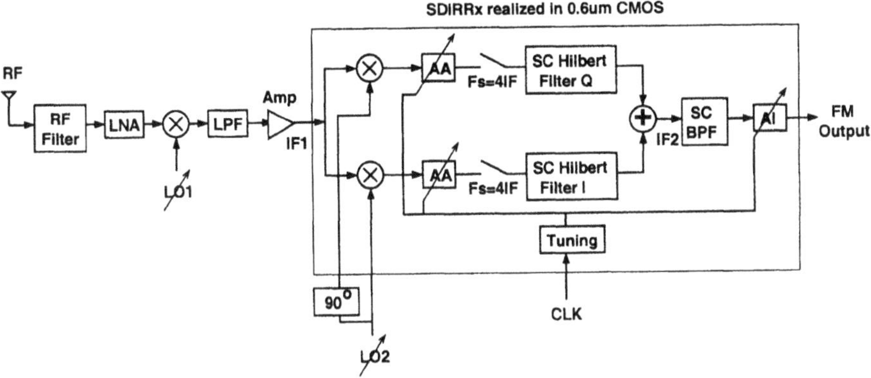

Figure 4.2: System architecture of an SDIRRx designed for cordless telephone applications.

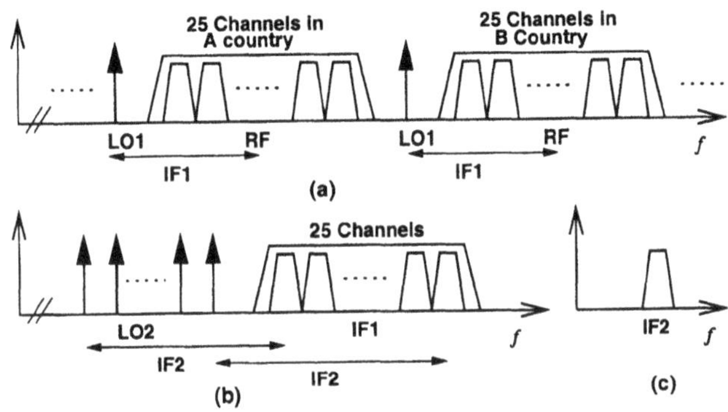

Figure 4.3: Spectra developed at RF, IF_1 and IF_2 stage.

mixer. Thus the first local oscillator performs rough tuning. Its tuning range is from 10 MHz to 40 MHz. Usually, this LO is not continuously tunable but can be switched from one band to another band. A lowpass filter, not a high-Q bandpass image rejection filter, is required at the first IF. This filter can be easily integrated.

The frequency down-conversion from the first IF to second IF is performed by the SDIRRx. The second local oscillator has a tuning range from 10 MHz to 10.48 MHz. It performs fine tuning. The second IF is 192 kHz. Output of the SDIRRx then passes through a bandpass filter (BPF) to perform channel selection. The BPF has a quality factor of 10, so can be easily implemented using SC technology. The discrete-time output signal is then converted back to continuous time form by an anti-imaging filter. The center frequency of the two anti-aliasing filters and the anti-image filter are tuned by an on-chip self-tuning circuit.

These circuit blocks are described in more detail in the subsequent subsections.

4.3.2 Mixer

The mixer topology used in this design is based on the linear region operation of MOS transistors [12, 13, 14], as shown in Figure 4.4. This mixer uses double balanced structure to cancel out the common-mode DC biasing signals and the first-order nonlinear dependence of transconductance on drain-source voltage.

The M1, M2, M3 and M4 in Figure 4.4 are biased in linear region. The input and LO signals are applied to the gates and drains respectively. With a perfect virtual ground the currents through these transistors are

$$I_{DS,1} = \beta_1 \left(V_{in}^{+} - V_{LO,DC} - V_{Tn1} - \frac{V_{LO}^{+} - V_{LO,DC}}{2} \right) (V_{LO}^{+} - V_{LO,DC}) \quad (4.2)$$

$$I_{DS,2} = \beta_2 \left(V_{in}^{-} - V_{LO,DC} - V_{Tn2} - \frac{V_{LO}^{-} - V_{LO,DC}}{2} \right) (V_{LO}^{-} - V_{LO,DC}) \quad (4.3)$$

$$I_{DS,3} = \beta_3 \left(V_{in}^{+} - V_{LO,DC} - V_{Tn3} - \frac{V_{LO}^{-} - V_{LO,DC}}{2} \right) (V_{LO}^{-} - V_{LO,DC}) \quad (4.4)$$

$$I_{DS,4} = \beta_4 \left(V_{in}^{-} - V_{LO,DC} - V_{Tn4} - \frac{V_{LO}^{+} - V_{LO,DC}}{2} \right) (V_{LO}^{+} - V_{LO,DC}), \quad (4.5)$$

where $\beta = \mu C_{ox} \frac{W}{L}$ and $V_{LO,DC}$ is DC level of the LO which equals the DC level at the virtual ground of the amplifier. Assuming perfect transistor matching, then the output is obtained by

$$\begin{aligned} V_o^{+} - V_o^{-} &= R_f(I_1 + I_2 - I_3 - I_4) \\ &= \beta R_f (V_{in}^{+} - V_{in}^{-})(V_{LO}^{+} - V_{LO}^{-}). \end{aligned} \quad (4.6)$$

Two factors suspected to limit the linearity of the mixer are the finite amplifier gain and mismatch.

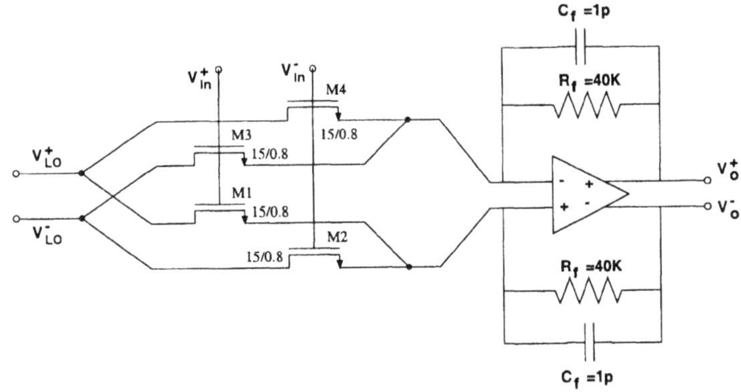

Figure 4.4: Circuit topology of the linear CMOS mixer.

The finite amplifier gain results a voltage change in the virtual ground. To see this effect, (4.2)-(4.5) should be modified to

$$\begin{aligned} I_{DS,1} &= \beta_1\left(V_{in}^+ - V_{LO,DC} + \Delta V - V_{Tn1} - \frac{V_{LO}^+ - V_{LO,DC} + \Delta V}{2}\right) \\ &\quad (V_{LO}^+ - V_{LO,DC} + \Delta V) \qquad (4.7) \\ I_{DS,2} &= \beta_2\left(V_{in}^- - V_{LO,DC} + \Delta V - V_{Tn2} - \frac{V_{LO}^- - V_{LO,DC} + \Delta V}{2}\right) \\ &\quad (V_{LO}^- - V_{LO,DC} + \Delta V) \qquad (4.8) \\ I_{DS,3} &= \beta_3\left(V_{in}^+ - V_{LO,DC} - \Delta V - V_{Tn3} - \frac{V_{LO}^- - V_{LO,DC} - \Delta V}{2}\right) \\ &\quad (V_{LO}^- - V_{LO,DC} - \Delta V) \qquad (4.9) \\ I_{DS,4} &= \beta_4\left(V_{in}^- - V_{LO,DC} - \Delta V - V_{Tn4} - \frac{V_{LO}^+ - V_{LO,DC} - \Delta V}{2}\right) \\ &\quad (V_{LO}^+ - V_{LO,DC} - \Delta V), \qquad (4.10) \end{aligned}$$

where $\Delta V = (V_o^+ - V_o^-)/2A$ is the voltage variation at virtual ground and A is the amplifier open-loop gain. Still assuming perfect matching, we have

$$\begin{aligned} V_o^+ - V_o^- &= \frac{R_f}{1+1/A}(I_1 + I_2 - I_3 - I_4) \\ &= \frac{\beta R_f}{1+1/A}\left[(V_{in}^+ - V_{in}^-)(V_{LO}^+ - V_{LO}^-) + 4\Delta V(V_{in,DC} - V_{LO,DC} - V_{Tn})\right] \qquad (4.11) \end{aligned}$$

where $V_{in,DC} = (V_{in}^{+} + V_{in}^{-})/2$. Substituting $\Delta V = (V_o^{+} - V_o^{-})/2A$ to the above equation, the output voltage is obtained as:

$$V_o^{+} - V_o^{-} = \frac{\beta R_f}{(1+\frac{1}{A})\left\{1 - 2\frac{\beta R_f}{1+A}(V_{in,DC} - V_{LO,DC} - V_{Tn})\right\}}(V_{in}^{+} - V_{in}^{-})(V_{LO}^{+} - V_{LO}^{-}). \tag{4.12}$$

The first term of (4.12) is a constant. Therefore the finite amplifier gain just affects the mixer gain but not linearity.

The mismatch between input transistors has two impacts. The first one is the appearance of residual DC-offset voltages caused by both β and V_T mismatches. These offset voltages either appear directly on the output of the mixer or result in feed-through of the RF and LO signal to the output caused by multiplication of these signals with the offset voltage. The second effect is the appearance of a quadratic V_{LO} component in the output signal:

$$V_o^{+} - V_o^{-} = \beta R_f (V_{in}^{+} - V_{in}^{-})(V_{LO}^{+} - V_{LO}^{-}) + \Delta\beta R_f (V_{LO}^{+} - V_{LO}^{-})^2, \tag{4.13}$$

caused by β mismatch. This explains why the RF signal is best applied to the gates of the modulating transistors. A quadratic V_{in} component at the output is highly unwanted since it results in a spurious signal at baseband. On the other hand, the quadratic V_{LO} signal results in an extra DC component at the output. AC coupling technique can be applied to connect the output of the mixer to next stage since the output is an IF signal.

DC biasing level of RF and LO ports must be carefully defined to avoid transistors M1-M4 being turned-off or entering saturation region. In the current design, $5V$ supply is used. The LO and RF DC levels are taken to be $2V$ and $3.5V$ respectively. The maximum LO signal can be applied is $8V_{pp}$ differential. The maximum RF signal that can be applied without turning off the modulating transistors is four times $V_{RF,DC} - V_{LO,DC} - V_T$, which gives $2.6V_{ptp}$ differential for the $0.6\mu m$ CMOS process used in the design. The W/L value of modulated transistors is 15/0.8, their drain-source conductance g_D is $14\,mS$. The feedback resistance and capacitance are 40 Ω and 1 pF respectively as shown in Figure 4.4.

Opamp design

The opamp used in this design is a fully balanced buffered single stage amplifier as shown in Figure 4.5. The topology is a folded-cascode operational transconductance amplifier with a capacitive compensation load. An output source follower performs buffering and level shifting. The opamp runs on a single 5V power supply. The output DC level is 2V. It is kept on this level by a common-mode feedback circuit shown in Figure 4.6(a). An on-chip resistor is used to generate biasing voltage as shown in Figure 4.6(a). This biasing circuit is shared by all the opamps used in mixers and continuous time filters (described in the subsequent subsection), to save silicon area.

For a 40 $k\Omega$ and a 1 pF load, the opamp has a frequency response shown in Figure 4.7. The DC gain is 70 dB, gain-bandwidth-product is 100 MHz, and the phase margin is 60°.

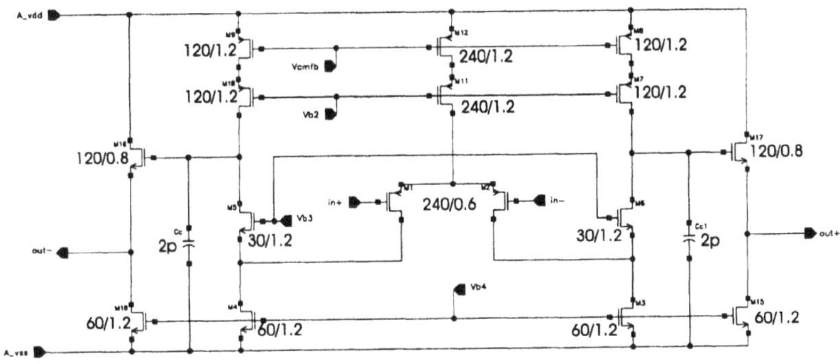

Figure 4.5: Circuit topology of the opamp used in the mixer.

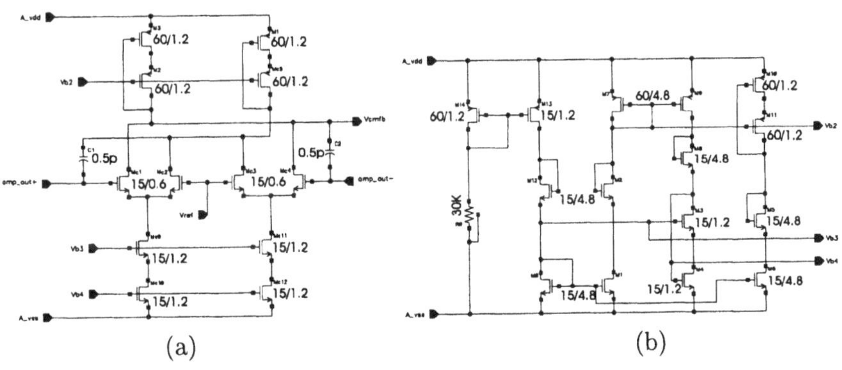

Figure 4.6: (a) Common-mode feedback and (c) bias circuitry of the opamp.

4.3.3 Anti-Aliasing and Anti-Imaging Filter

The specifications of anti-aliasing and anti-imaging filters (AAF and AIF) in this receiver are exactly the same. The signal band is from 90 kHz to 110 kHz,

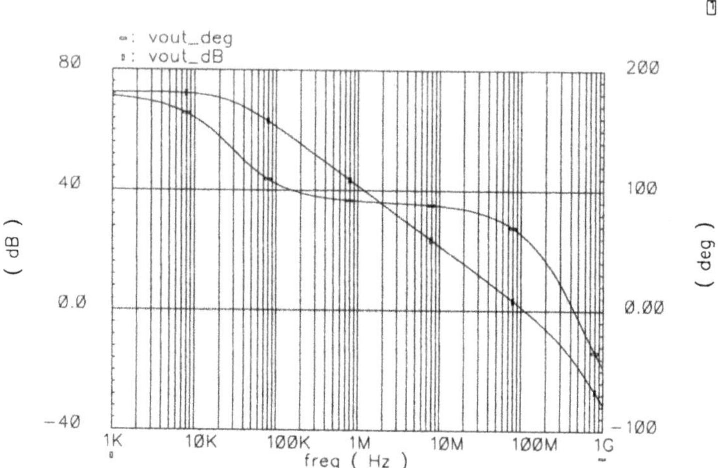

Figure 4.7: Frequency response of the opamp.

and the sampling frequency is 400 kHz. As to be described later, the signal will pass through an SC bandpass filter with a stop band from 0 to 70 kHz and from 130 kHz to 200 kHz. This means that out-of-band energy from 130 kHz to 270 kHz will be filtered by the bandpass filter, and the AAF or AIF just needs to reject signals above 270 kHz. These specifications are depicted in Figure 4.8.

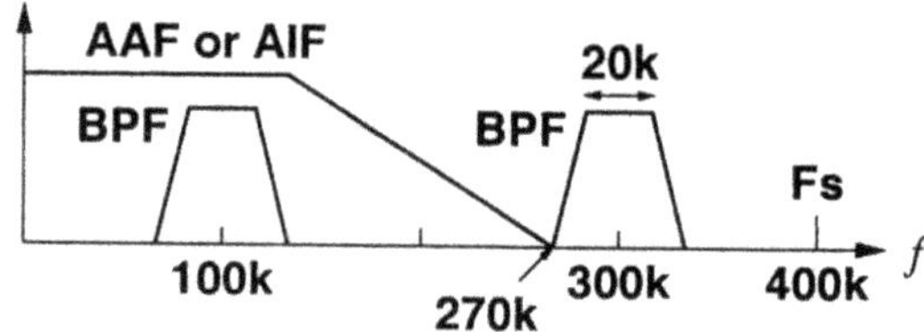

Figure 4.8: The anti-aliasing and anti-imaging filter specifications.

With the pass-band frequency of 130 kHz, the stop-band frequency of 270 kHz, the pass-band ripple of 0.5 dB, and the stop-band attenuation of 50 dB, a fourth order elliptic lowpass filter prototype is obtained for the AAF and AIF. Its s-domain transfer function is given by:

$$H(s) = K\frac{(s^2 + b_0)(s^2 + b_1)}{(s^2 + a_{00}s + a_{01})(s^2 + a_{10}s + a_{11})} \tag{4.14}$$

where $K = 3.1637 \times 10^{-3}$, $b_0 = 1.7453 \times 10^{13}$, $b_1 = 3.2845 \times 10^{12}$, $a_{00} = 7.203 \times 10^5$, $a_{01} = 2.7094 \times 10^{11}$, $a_{10} = 2.4889 \times 10^5$ and $a_{11} = 7.0901 \times 10^{11}$.

MOSFET-C filter technique [15] was chosen to realize the AAF and AIF for the sake of better linearity (compared to $g_m - C$ technique) and simplicity. Frequency tuning is required and has been realized in this design by an on-chip SC control circuit which is described separately in the next subsection.

In order to have more flexibility to optimise dynamic range, facilitate the requirements of tuning and ease layout, a cascade-of-biquads topology was chosen. Standard design procedures have been followed to perform pole-zero pairing, section ordering and gain partitioning [16]. Zeros and poles corresponding to coefficients b_0, a_{00} and a_{01} are assigned to the first biquad which has a lower Q value, and those corresponding to coefficients b_1, a_{10} and a_{11} are assigned to the second biquad which has a higher Q value. To maximise the dynamic range, the constant K is divided to 0.03888 and 0.081371 for the low-Q and high-Q biquads respectively. For the same purpose, the high-Q biquad should be placed before the low-Q biquad.

A signal flow diagram of both the high-Q and low-Q biquads is shown in Figure 4.9. The biquads have a general transfer function of

$$H(s) = -\frac{k_1(s^2 + \frac{k_0}{k_1})}{s^2 + \frac{\omega_o}{Q}s + \omega_o^2}. \tag{4.15}$$

An active-RC circuit to realize this transfer function is shown in Figure 4.10(b). By using balanced MOS transistors to approximate linear resistors [17], we obtain the MOSFET-C biquad circuit as shown in Figure 4.10(a). All the transistors have their sources connected to virtual ground, and their gates connected to a control voltage V_{ctrl} which is the output of an SC control circuit. The fully balanced structure of this circuit gives a good approximation of a MOSFET to a linear resistor. In terms of component values, the biquad shown in Figure 4.10(a) has a transfer function of:

$$H(s) = -\frac{C_1}{C_B}\frac{s^2 + \frac{G_1G_2}{C_1C_A}}{s^2 + \frac{G_4}{C_B}s + \frac{G_2G_3}{C_AC_B}}. \tag{4.16}$$

Comparing (4.15) with (4.16), four design equations are obtained:

$$\frac{C_1}{C_B} = k_1, \quad \frac{G_1G_2}{C_1C_A} = \frac{k_0}{k_1}, \quad \frac{G_4}{C_B} = \frac{\omega_o}{Q}, \quad \text{and} \quad \frac{G_2G_3}{C_AC_B} = \omega_o^2, \tag{4.17}$$

for seven variables C_1, C_A, C_B, G_1, G_2, G_3 and G_4. C_1 is first set to be $0.1pF$ as the unit capacitor. Some degree of freedom is allowed in adjusting

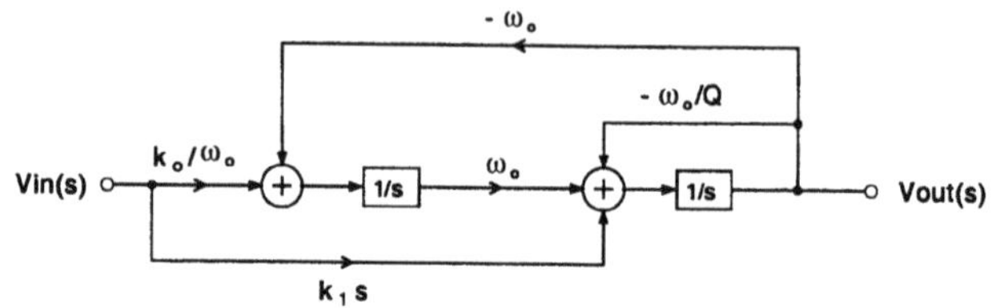

Figure 4.9: A signal flow diagram of a biquad.

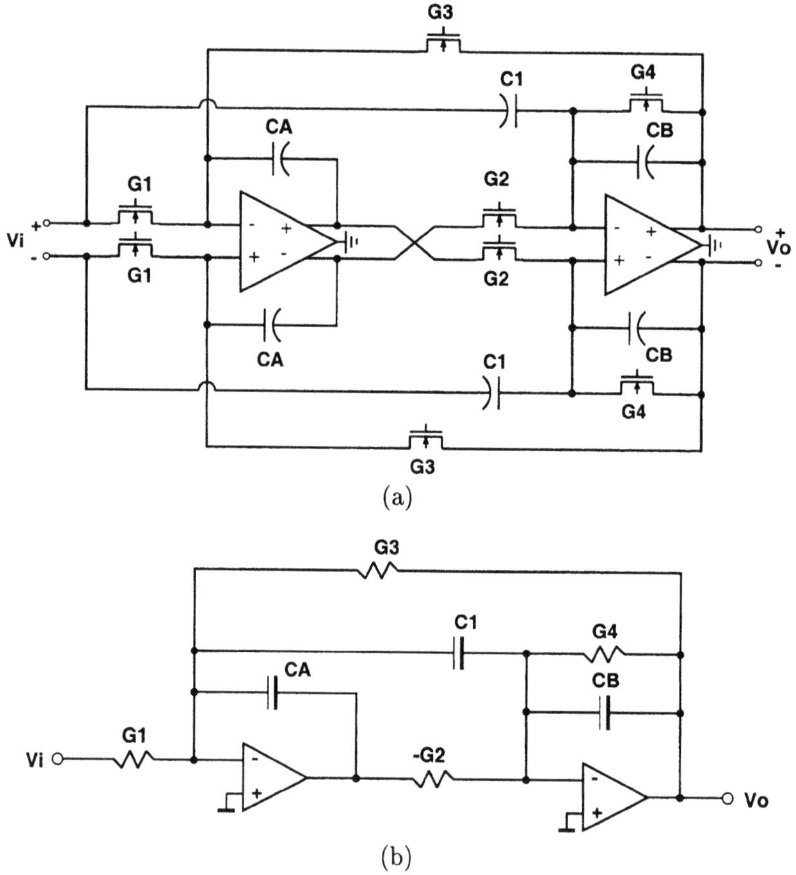

(a)

(b)

Figure 4.10: (a) A MOSFET-C biquad section. (b) The equivalent active-RC half circuit.

these component values for optimising the dynamic range of the output of the first opamp. In the current design, the capacitor values and transistor sizes for both high-Q and low-Q biquads are listed in Table 4.2. Note that the sizing of transistors is closely related to the tuning mechanism and will be explained in the subsequent subsection.

The frequency response of the designed filter is shown in Figure 4.11.

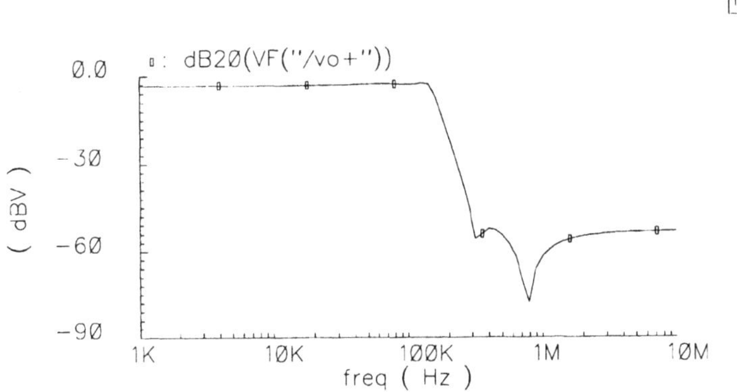

Figure 4.11: Frequency response of the 4th order elliptic MOSFET-C lowpass filter.

Table 4.2: Component values of MOSFET-C biquads.

	C_1	C_A	C_B	$(L/W)_1$	$(L/W)_2$	$(L/W)_3$	$(L/W)_4$
1	0.1 pF	1.25 pF	1.25 pF	49.2/0.8	18.6/0.8	18.6/0.8	62.8/0.8
2	0.1 pF	5 pF	2.5 pF	5.7/0.8	7.2/0.8	14.4/0.8	10.4/0.8

Biquad 1 and 2 are the high-Q and low-Q biquads respectively.

4.3.4 SC Control Circuit for MOSFET-C Filter Tuning

Tuning of the AAF and AIF is carried out by means of the SC circuit shown in Figure 4.12, a modified version of an SC frequency control circuit described in [18].

The first opamp whose negative input terminal is connected to a voltage divider is used to generate balanced DC reference voltages V_B and $-V_B$ for the next stage. The clock phasing is arranged so that the SC block emulates a negative resistor, whose equivalent resistance R_{SC} is given by the following

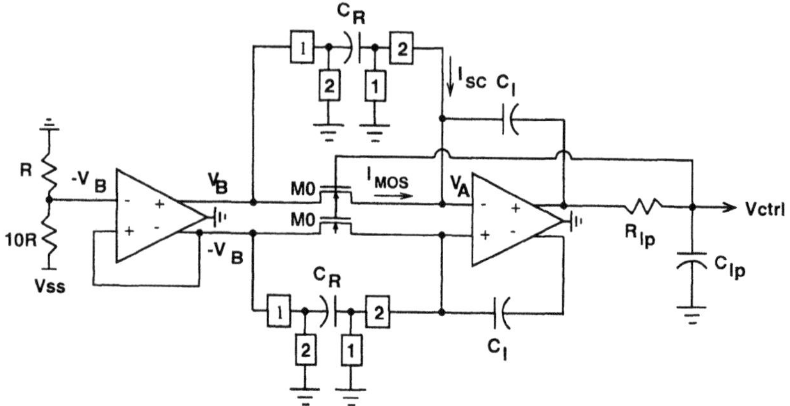

Figure 4.12: The tuning circuit of the MOSFET-C filters.

formula:

$$R_{SC} = -\frac{1}{f_{clk}C_R}. \tag{4.18}$$

The conductance G_0 formed by balanced transistor M_0 is $\mu C_{ox}(W/L)_0(V_{ctrl} - V_T)$. One can observe that the feedback of output V_{ctrl} to the gate of M_0 is negative. In the steady-state, this negative feedback mechanism will force the current entering the integrating capacitor C_I to be zero, i.e., $I_{SC} + I_{MOS} = 0$. Hence,

$$(V_B - V_A)G_0 + \frac{V_B - V_A}{R_{SC}} = 0. \tag{4.19}$$

Substituting (4.18) to (4.19), we obtain the tuning equation

$$\frac{G_0}{C_R} = f_{clk}, \tag{4.20}$$

which means that the time constant C_R/G_o is controlled by the clock frequency of the SC control circuit. If the dimensions of transistors and capacitors in the tuning circuit (the master) are well matched to those in the main filters (the slaves), then those design parameters G_i/C_j in (4.17) are proportional to f_{clk}, which is a precise parameter. Therefore, from (4.20) and (4.16), we have the biquad's transfer function in terms of relative component values:

$$H(s) = -\frac{C_1}{C_B}\frac{s^2 + \frac{G_1/G_0}{C_1/C_R}\frac{G_2/G_0}{C_A/C_R}f_{clk}^2}{s^2 + s\frac{G_4/G_0}{C_B/C_R}f_{clk} + \frac{G_2/G_0}{C_A/C_R}\frac{G_3/G_0}{C_B/C_R}f_{clk}^2}, \tag{4.21}$$

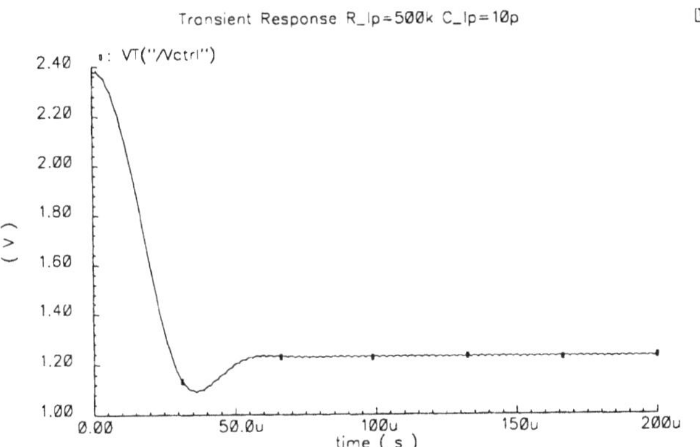

Figure 4.13: Transient response of the SC tuning circuit of MOSFET-C filter.

where the conductance ratio between two transistors equals the ratio between their (L/W) value. From (4.21), it can be found that the the 3-dB cut-off frequency ω_o of the biquad depends linearly on f_{clk}. But the Q-factor is fixed. Note that the opamps in the master must also be well matched to those in slaves. An advantage of this tuning circuit is that the reference clock signal can be easily programmable, and thus allowing tunability over a wide bandwidth.

In the current design, the clock frequency f_{clk} is 400 kHz, the dimension of M_0 is $L/W = 75/0.8$, the value of C_R and C_I are 0.64 pF and 1.2 pF respectively. Resistor R_{lp} and capacitor C_{lp} form the low-pass filter used to remove the periodic component at clock frequency existed in the output of the second opamp. This periodic component exists since the current injected to C_I from R_{SC} is a periodic signal at the clock frequency. The time-constant $R_{lp}C_{lp}$ is set to be 5 μs. Off-chip components are used for R_{lp} and C_{lp} due to their large value. The circuit runs at 5V supply, and the common-mode voltage is set to 2V, exactly the same as those in the main filter. The transient response of the tuning circuit is shown in Figure 4.13.

4.3.5 SC Hilbert Transformer

After the anti-aliasing filter, IF signals at I and Q paths are then sampled at 400 kHz by the SC Hilbert transformer pair which performs 90° phase shift. The IF signals are centered at 100 kHz with a bandwidth of 20 kHz. The Hilbert transformer is designed to have a bandwidth of 30 kHz, and an image rejection ratio of over 60 dB. Following the procedures described in Chapter

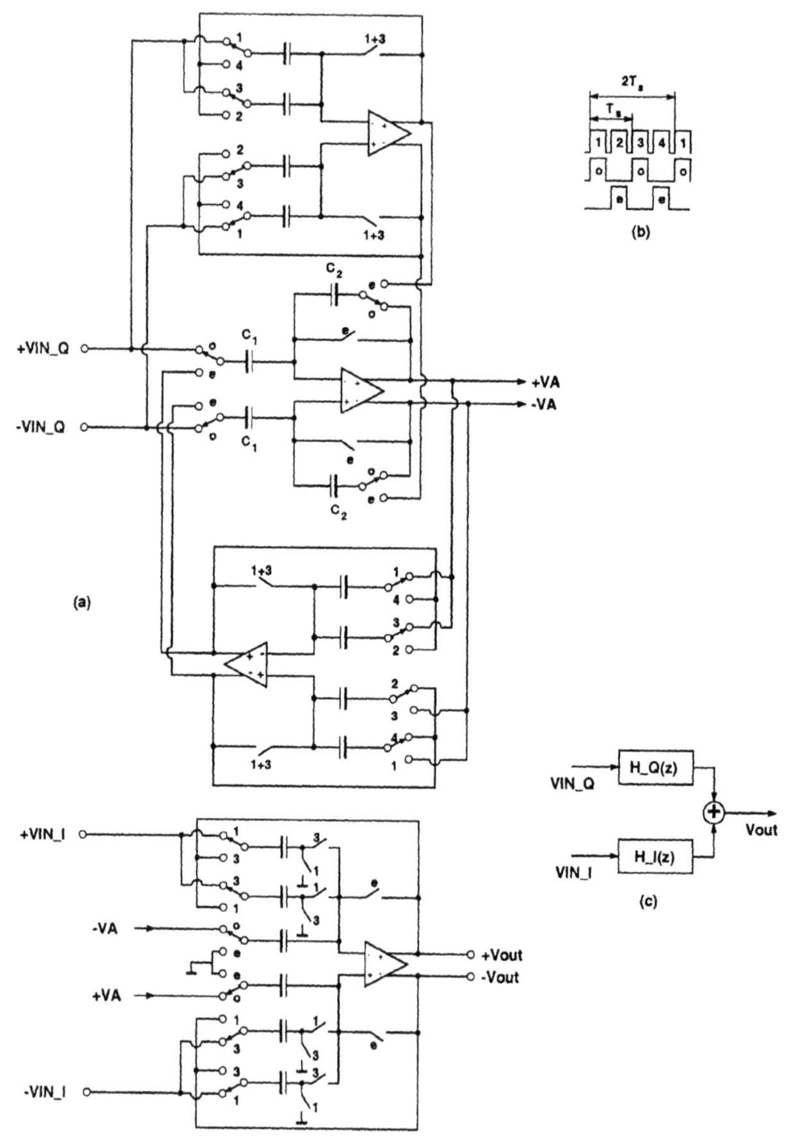

Figure 4.14: (a) Fully differential SC Hilbert transformer circuit embedded with an adder. (b) Its phase diagram. (c) Its equivalent block diagram. $C_1/C_2 = 1/3$, and others are unit capacitors.

three, we obtain the transfer functions of an IIR Hilbert transformer pair as

$$H_I(z) = z^{-1}, \tag{4.22}$$

$$H_Q(z) = \frac{z^{-2} - 0.338019}{1 - 0.338019z^{-2}}, \tag{4.23}$$

which has a stop-band attenuation of 67 dB. To reduce the capacitance spread, the filter coefficients of $H_Q(z)$ are truncated to the nearest integer as:

$$H_Q(z) = \frac{3z^{-2} - 1}{3 - z^{-2}}. \tag{4.24}$$

This change has little impact on the filter response.

A fully differential SC circuit realizing the Hilbert transformer is shown in Figure 4.14(a). Poly-phase circuits have been adopted to implement the delay functions of $z^{3/2}$ and z^{-1}. The phase diagram is shown in Figure 4.14(b). They are externally supplied in the current design. Thanks to the poly-phase technique, only four amplifiers are used here, while a traditional two phase circuit requires nine amplifiers (see Chapter three). Therefore a large amount of silicon area and power is saved.

Figure 4.14(c) gives the equivalent block diagram of the circuit shown in Figure 4.14(a). Note that the summing circuit is embedded in filter-I. The frequency response of the circuit for the wanted signal and image are shown in Figure 4.15. The simulated IRR of the SC Hilbert transformer is over 64 dB in the 20 kHz signal band.

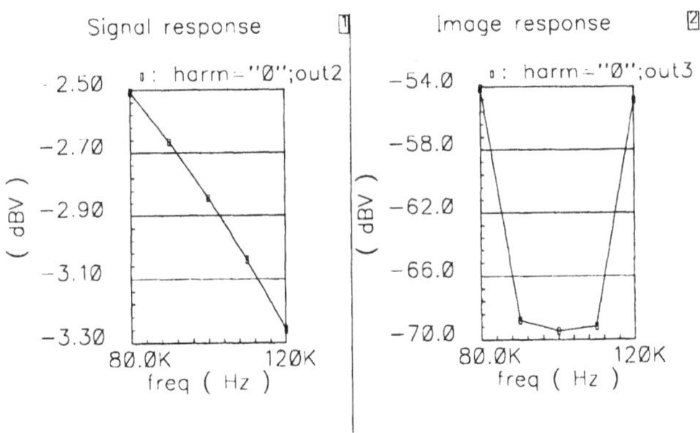

Figure 4.15: Frequency response of the SC Hilbert transformer for the wanted signal and the image interferer.

Operational Transconductance Amplifier

Fully differential operational transconductance amplifiers (OTA) are used in the SC circuit. Folded-cascode topology is employed as shown in Figure 4.16(a). The bias circuit shown in Figure 4.16(b) is shared for all OTAs. A standard SC common feedback circuit is employed as shown in Figure 4.16(c). It runs at the same sampling frequency as the SC transformer. It samples the difference between the common mode voltage of the OTA's outputs and the reference voltage, and adds this difference to the bias voltage V_{b4}:

$$V_{cmfb} = V_{b4} + (V_{cm} - V_{ref}), \tag{4.25}$$

where $V_{cm} = (V_o^+ + V_0^-)/2$. The V_{cmfb} is connected to bias transistors M_3, M_4 and M_7 of the OTA to form a negative feed back loop.

The OTA runs at single 5 V supply, and the common-mode voltage is set at 2 V. For a 1 pF load, the OTA has a frequency response as shown in Figure 4.17. The DC gain is 70 dB, gain-bandwidth product is 120 MHz, and the phase margin is 60°.

Switch

CMOS switches are used for minimising clock-feedthrough and charge injection effects. The pMOS and nMOS of all switches have a dimension of $W/L = 3\mu/0.6\mu$. The on-resistances of the pMOS and nMOS as a function of input voltage level are shown in Figure 4.18.

4.3.6 SC Bandpass Filter

The channel selection bandpass filter has a center frequency of 100 kHz, a passband from 90 kHz to 110 kHz. The lower and upper stop band frequency are 70 kHz and 130 kHz respectively. The pass band ripple and stop band attenuation are 3 dB and 40 dB respectively.

The topology of cascaded biquads is chosen. By using Xfilter [19], a sixth order elliptic filter was designed. It consists of three cascaded biquads, having transfer function of

$$H_1(z) = -0.10124\frac{z^2 + 1.0095z + 1}{z^2 + 1.10220}, \tag{4.26}$$

$$H_2(z) = -0.10643\frac{z^2 - 2z + 1}{z^2 - 0.2917z + 1.0457}, \tag{4.27}$$

$$H_3(z) = -0.4058\frac{z^2 - 1.0095z + 1}{z^2 + 0.2917z + 1.0457} \tag{4.28}$$

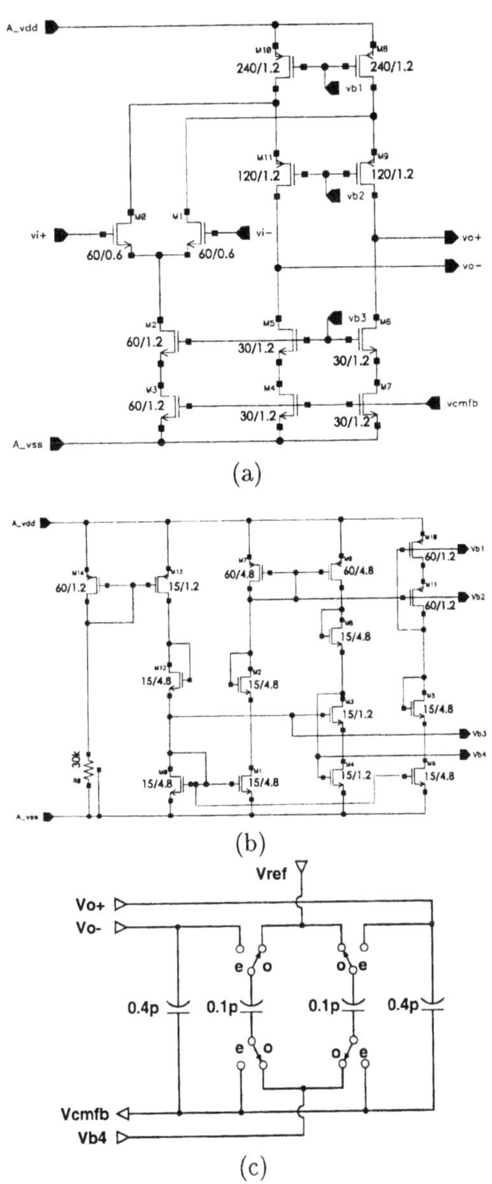

Figure 4.16: (a) The OTA used in the SC circuits. (b) The bias circuit. (c) The common-mode feedback circuit.

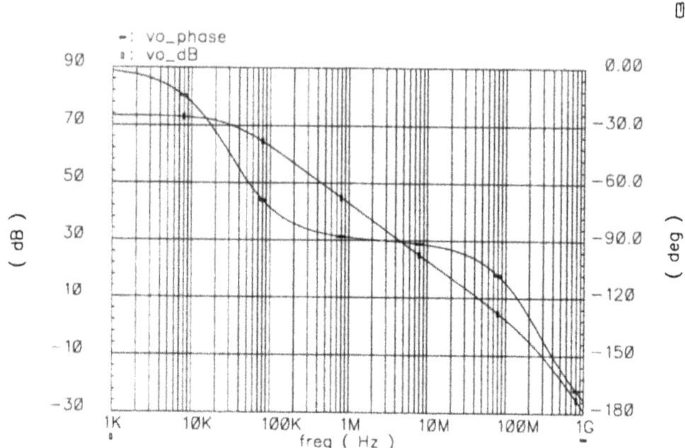

Figure 4.17: Frequency response of the OTA.

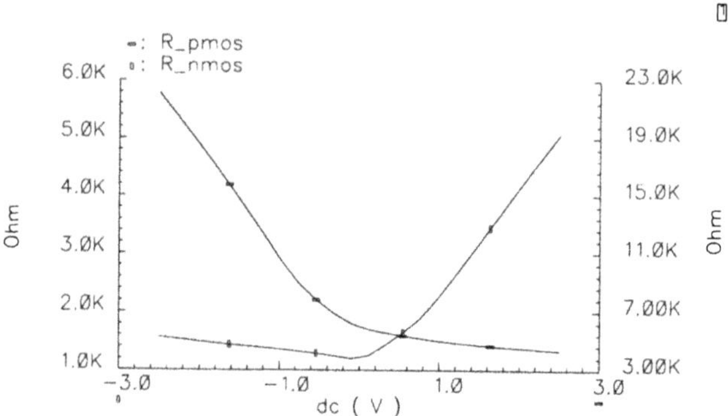

Figure 4.18: On resistance of the pMOS and nMOS as a function of input voltage.

respectively.

The three SC biquads are shown in Figure 4.19. Their capacitor values are listed in Table 4.3. The number of OTAs, capacitors and switches employed in this design are 6, 27 and 41 respectively. Total capacitance is about 20 pF with the unit capacitance of 0.2 pF. The capacitance spread is only 15. All the OTAs and switches used in the SC BPF are identical to those used in SC Hilbert transformers.

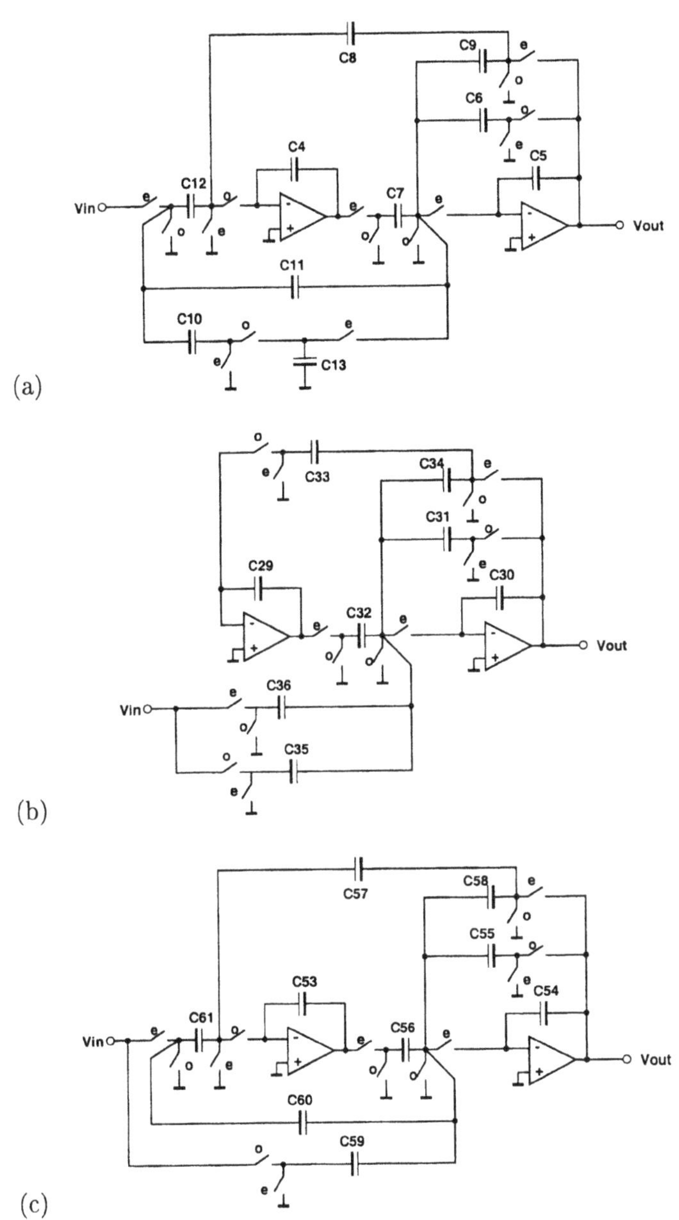

Figure 4.19: (a-c) The first, second and third biquad of the 6th-order switched-capacitor bandpass filter.

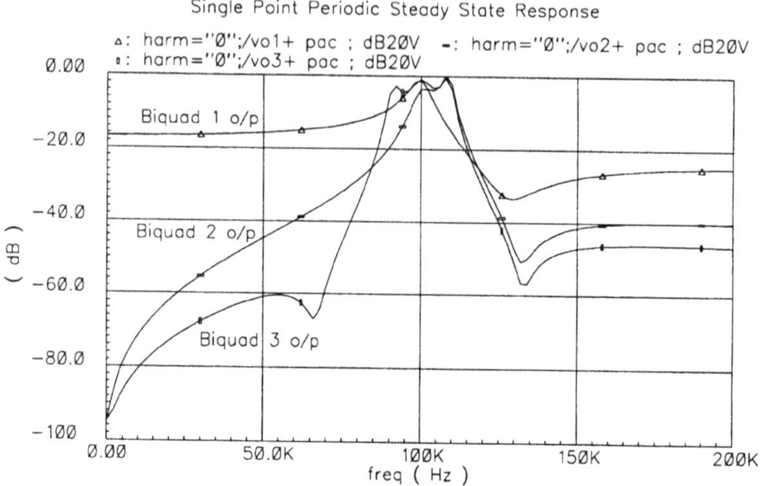

Figure 4.20: Frequency response of the SC bandpass filter.

Table 4.3: Capacitor values of the SC bandpass filter.

Biquad I	Biquad II	Biquad III
$C_{11} = C_{12} = 0.2$	$C_{29} = C_{35} = C_{36} = 0.2$	$C_{59} = C_{60} = C_{61} = 0.2$
$C_{10} = C_{13} = 0.4$	$C_{30} = C_{31} = 0.983$	$C_{53} = 0.681$
$C_4 = 0.986$	$C_{32} = 2.888$	$C_{54} = C_{55} = 0.258$
$C_5 = C_6 = 1.089$	$C_{33} = 0.304$	$C_{56} = 0.675$
$C_7 = 2.967$	$C_{34} = 0.897$	$C_{37} = 0.873$
$C_8 = 1.38$		$C_{58} = 0.235$
$C_9 = 0.887$		

All capacitors are in unit of pF.

The frequency response of the SC BPF is shown in Figure 4.20. The output levels of all biquad sections are designed to around 0 dB in the pass band for maximising the dynamic range of the filter.

Layout

The SDIRRx was realized in a 0.6 μm double-poly double-metal CMOS technology. The micro-photograph of the chip is shown in Figure 4.21. The size of the chip is $2.5mm \times 2.8mm$. The active circuit area is only $1mm \times 1.5mm$.

The chip was packed in a 28-pin dual-in-line package, while the actually used pin number is 25.

The floor-planning of the chip is also shown in Figure 4.21. I and Q mixers, MOSFET-C anti-aliasing and anti-imaging filters and their tuning circuit are placed in the left and occupy about two-third of the active area. The SC Hilbert transformer and the SC bandpass filter are placed together and occupy about one-third of the active area. The only digital part of the chip is the externally supplied clock lines. These clock lines are put at the right edge of the active circuit area.

Figure 4.21: Chip micro-photograph.

4.4 Chip Simulation Results

Post-layout simulations were conducted to test the image rejection performance of the SDIRRx chip. Results are presented in this section.

To test the receiver's response to the desired signal, a sinusoidal input was swept from 10.085 MHz to 10.115 MHz with a step of 5 kHz, while the LO frequency is fixed at 10 MHz. The corresponding output frequency varies from 85 kHz to 115 kHz, which is within the specified band of interest. Transient simulations from 0 to 300 μs were conducted for different input frequencies. The magnitude of the output signal was computed by taking a 256-point, rectangular-window DFT on a section of its waveform, from 100 μs to 300 μs, which corresponds to a frequency step of 5 kHz. Note that the first 100 μs data are not used for the tuning circuit and common-mode feedback circuits to settle. To test the receiver's response to the image interferer, the sinusoidal input was swept from 9.885 MHz to 9.915 MHz. The output magnitude was computed by the same DFT process mentioned above.

Table 4.4 gives the results of the simulations, where the magnitudes of RF input and LO were set to 0.4 V_{pp} (differential) and 1.376 V_{pp} (differential) respectively. Mismatches are not considered. The results are also depicted in Figure 4.22. The simulated image rejection ratio of the SDIRRx is better than 47 dB within the band of interest.

Table 4.4: Simulated responses of the SDIRRx to desired signal and image interferer.

	input frequency	output frequency	Output magnitude
Wanted Signal	10.085 MHz	85 kHz	-0.5674 dB
	10.090 MHz	90 kHz	-0.4709 dB
	10.095 MHz	95 kHz	-0.3711 dB
	10.100 MHz	100 kHz	-0.4663 dB
	10.105 MHz	105 kHz	-0.2561 dB
	10.110 MHz	110 kHz	-0.251 dB
	10.115 MHz	115 kHz	-0.2747 dB
Image Interferer	9.915 MHz	85 kHz	-50.12 dB
	9.910 MHz	90 kHz	-52.49 dB
	9.905 MHz	95 kHz	-50.83 dB
	9.900 MHz	100 kHz	-49.82 dB
	9.895 MHz	105 kHz	-47.5 dB
	9.890 MHz	110 kHz	-47.24 dB
	9.885 MHz	115 kHz	-48.79 dB

4.5 Experimental Evaluation

4.5.1 Measurement Setup

The laboratory set up for evaluating the SDIRRx is presented in Figure 4.23. A two face test PCB is used.

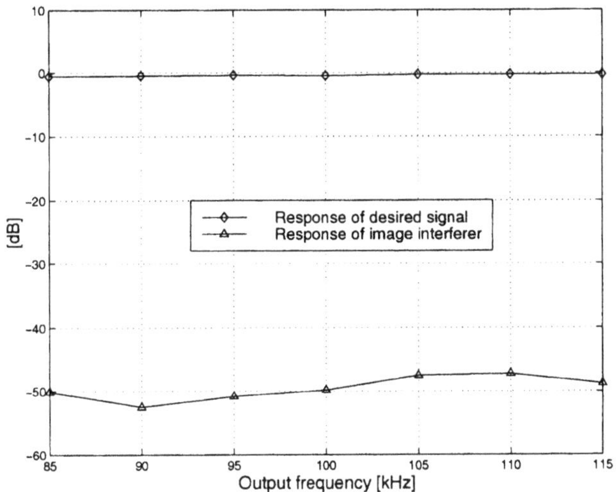

Figure 4.22: Simulated image rejection performance of the SDIRRx.

Sets of 100 nF, 1 μF and 100 μF capacitors are used to decouple power supply AVDD and VCM from ground (AVSS). All the amplifiers used in the SDIRRx are internally biased. Six clock signals are externally supplied by using a digital function generator (HP8166A). A 50 $k\Omega$ resistor and a 100 pF capacitor, constructed as a lowpass filter, are used for the tuning circuit of the MOSFET-C filters.

RF and LO signals are converted to differential form before injecting to the chip by using transformers. Balance of the differential signals is ensured by tying the center tap of the secondary wind of the transformer to V_{cm}. 10 MHz sinusoidal LO signal is generated from HP3245. A lowpass filter with cut-off frequency of 15 MHz is used to purify the LO input. The quadrature component of LO is generated by an RC polyphase network as shown in Figure 4.23.

Differential IF output and intermediate outputs from AAF and SC bandpass filter are converted back to single-end form for measurement. Due to its frequency-shifting nature, the receiver is characterised by using separate signal generation and detection instruments. The RF signal is swept by using spectrum analyser HP3588, and the baseband output spectrum is observed from spectrum analyser HP8595. These instruments are connected through HPIB and synchronised under computer control.

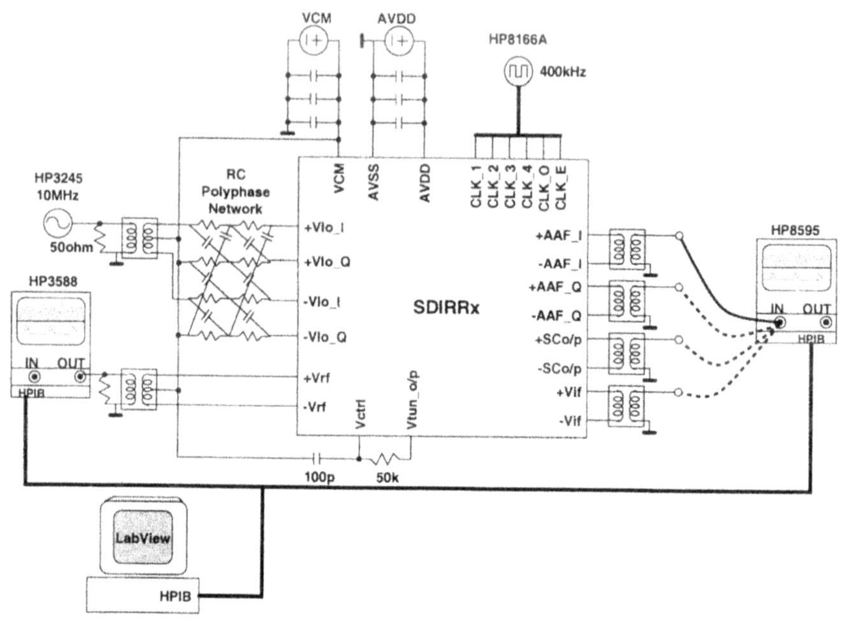

Figure 4.23: The measurement setup.

4.5.2 Measurement Results

Due to a fault in the fabrication, the SC part of the chip, including the SC Hilbert transformer and SC bandpass filter (except the SC control circuit for MOSFET tuning), does not work. Experimental results discussed in this section will be therefore concentrated on the tunable front-end of the SDIRRx.

The tunable front-end occupies an area of 0.56 mm^2. Figure 4.24 shows the micro-photograph of tunable front-end. Current consumption of less than 5 mA from a single 5V supply was measured on 10 samples.

To measure the frequency response of the self-tuned MOSFET-C filter, the mixer gain was adjusted to 1, the tuning clock was set at 400 kHz, and the input was swept from 10.01 MHz to 10.5 MHz. The measured output response is displayed in Figure 4.25. It can be observed that the filter has a 3-dB cutoff frequency of 130 kHz and 50 dB attenuation at 270 kHz. This matches very well with the designed response (dotted curve in the figure). A small tone at 400 kHz was also observed. This is due to the interference coupling from the clock signal.

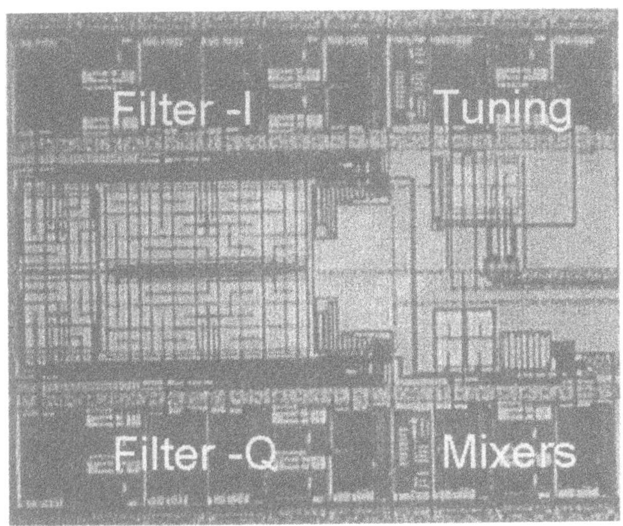

Figure 4.24: Micro-photograph of the tunable front-end.

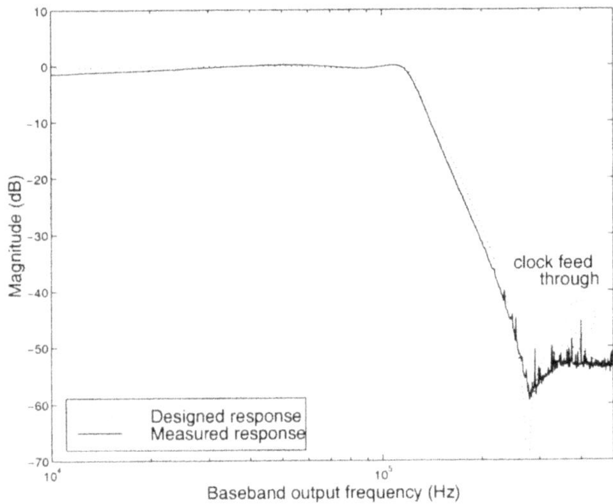

Figure 4.25: Magnitude response representing the combined functions of frequency down conversion and filtering of the receiver, with LO at 10 MHz and $f_{clk} = 400kHz$.

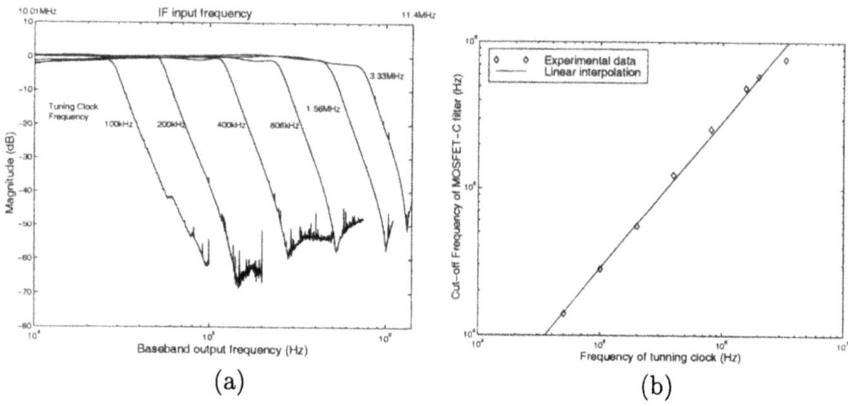

(a) (b)

Figure 4.26: (a)Magnitude responses representing the combined functions of frequency down conversion and filtering of the receiver under different f_{clk}, with LO at 10 MHz. (b) 3-dB cutoff frequency of the MOSFET-C filter as a function of the tuning clock frequency.

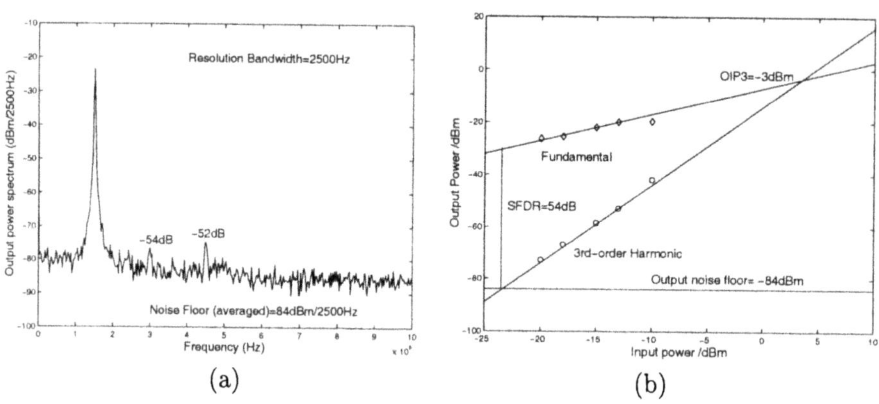

(a) (b)

Figure 4.27: (a)Output frequency spectrum for a 10.15 MHz sinusoidal input, where LO and tuning clock were set at 10 MHz and 3.33 MHz respectively. (b) Measured 3rd-order intercept point.

The tunable front-end was then tested with different tuning clock frequency. The measured output responses with respective to tuning clock frequencies of 100 kHz, 200 kHz, 400 kHz, 806 kHz, 1.56 MHz and 3.33 MHz are plotted in Figure 4.26(a). It is observed that the 3-dB cut-off frequency of the filter varies linearly with the tuning clock frequency, as depicted in Figure 4.26(b). The tunable range of the cut-off frequency is from 15 kHz to 800 kHz, a range of almost two orders of magnitude. The upper limit is set up by the supply voltage because higher tuning frequency leads to higher V_{ctrl} which can not exceed V_{dd}. In the lower limit region, the slope of the amplitude response in the transition region is degraded. This is due to the fact that MOS transistors used in the filter are no longer properly emulating linear resistors when they enter the saturation region for low gate control voltage V_{ctrl}.

To examine the single-tone response of the front-end, a 10.15 MHz sinusoidal RF input was applied, with the LO set at 10 MHz. Besides, the tuning clock was set at 3.33 MHz to accommodate the second and third harmonics in the passband. The measured output spectrum is displayed in Figure 4.27(a). It is observed that the second and third harmonics are 54 dB and 52 dB below the fundamental respectively. The third harmonic is dominant because even harmonics are suppressed by the fully differential circuit architecture. Output noise level of -84 dBm for the resolution bandwidth of 2500 Hz was observed which corresponds to -118 dBm/Hz.

To examine the linearity, the third-order intercept point of the tunable front-end was measured and the result is shown in Figure 4.27(b). It is observed that the output-referred 3rd-order intercept point is -3 dBm, and the spurious-free dynamic range (SFDR) is 54 dB. The front-end exhibits a moderate linearity. This is mainly due to the linearity limitation of the transistors used the MOSFET-C filters. The balanced structure can not completely cancel the non-linearity of the transistors.

The experimental results of the front-end are summarised in Table 4.5.

4.6 Summary

In this chapter, a sampled-data image rejection receiver architecture has been proposed. The novelty of the receiver lies in the use of switched-capacitor Hilbert transformer as the 90° phase shifter. Phase shifting performed in this way is more accurate and has wider bandwidth than traditional approaches.

A receiver designed using this architecture for cordless telephone applications, was implemented in a 0.6 μm CMOS technology. The chip includes I and Q mixers, self-tuned MOSFET-C anti-aliasing and anti-imaging filters, an SC Hilbert transformer pair and an SC bandpass filter. A novel frequency con-

Table 4.5: Performance parameters of the tunable front-end.

IF input frequency range	0 - 50 MHz
MOSFET-C filter type	4th order elliptic lowpass
Cutoff frequency (Tuning clock @ 400 kHz)	130 kHz
Stop-band attenuation	50 dB @ 270 kHz
Tunable range (Cut-off frequency)	15 kHz - 800 kHz
Gain (LO @ $500mV_{pp}$)	0 dB
IP3 (Output-referred)	-3 dBm
Spurious Free Dynamic Range	54 dB
Power supply	5 V
Current consumption	5 mA
Active area	0.56 mm^2
Technology	0.6 μm CMOS

trol circuit for the MOSFET-C filters has been used. The cut-off frequency of the filters is precisely controlled by an external clock signal. Chip simulations show that over 47 dB of image rejection ratio can be achieved.

The chip was measured and characterised in the laboratory. It was not able to examine the image rejection performance of the receiver due to a defeat in the SC parts of the chip. However, the remaining parts of the chip worked properly. The self-tuning circuit was proven. It was found that, by changing the controlling clock frequency, the MOSFET-C filter had a very wide tuning range, from 15 kHz to 800 kHz, a range of almost two orders of magnitude.

References

[1] T. Okanobu, H. Tomiyama, and H. Arimoto, "Advanced low voltage single chip radio IC," *IEEE Trans. Consumer Electronics*, vol. 38, no. 3, pp. 465–475, August 1992.

[2] M. D. McDonald, "A 2.5GHz BiCMOS image-reject front end," in *Digest of Technical Papers, IEEE Int. Solid-State Circuit Conference*, 1993, pp. 144–145.

[3] Werner Baumberger, "A single-chip image rejecting receiver for the 2.44 GHz band using commercial GaAs-MESFET-technology," *IEEE J. Solid-State Circuits*, vol. 29, no. 10, pp. 1244–1249, Oct. 1994.

[4] D. Pache, J.M. Fournier, G. Billiot, and P. Senn, "An improved 3V 2GHz BiCMOS image reject mixer IC," in *Proc. IEEE Custom Integrated Circuits Conference*, 1995, pp. 95–98.

[5] K.P. Pun, J.E. Franca, and C. Azeredo Leme, "Basic principles and new solutions for analog sampled-data image rejection mixers," in *Proc. IEEE Int. Conference on Electronics, Circuits and Systems*, Lisbon, Portugal, Sept. 1998, vol. 3, pp. 165–168.

[6] J. Tang and D. Kasperkovitz, "A 0.9-2.2GHz monolithic quadrature mixer oscillator for direct-conversion satellite receivers," in *Digest of Technical Papers, IEEE Int. Solid-State Circuit Conference*, Feb. 1997, pp. 88–89.

[7] M. Steyaert and R. Roovers, "A 1-GHz single-chip quadrature modulator," *IEEE J. Solid-State Ciruits*, vol. 27, no. 8, pp. 1194–1197, Aug 1992.

[8] T.D. Stetzler, I.G. Post, J.H. Havens, and M. Koyama, "A 2.7-4.5V single chip GSM transceiver RF integrated circuit," *IEEE J. Solid-State Circuits*, vol. 30, no. 12, pp. 1421–1429, Dec 1995.

[9] S. Levantino, C. Samori, M. Banu, J. Glas, and V. Boccuzzi, "A CMOS IF sampling circuit with reduced aliasing for wireless applications," in *Dig. of Tech. Papers, IEEE Int. Solid-State Circuits Conference*, 2002, pp. 404–405.

[10] European Telecommunications Standards Institute, *Technical charateristics, test conditions and methods of measurement for radio aspects of cordless telephones CT1*, 1994.

[11] Motorola, *Universal Cordless Telephone Subsystem IC: MC13111*, 1996.

[12] B.S. Song, "CMOS RF circuits for data communications applications," *IEEE J. Solid-State Circuits*, vol. SC-21, no. 2, pp. 310–317, April 1986.

[13] Y.P. Tsividis, *Operation and modeling of the MOS transistor*, McGrawHill, 1987.

[14] J. Crols and M. Steyaert, "A 1.5GHz highly linear CMOS down conversion mixer," *IEEE J. Solid-State Circuits*, vol. 30, no. 7, pp. 736–742, July 1995.

[15] Yannis P. Tsividis, "Integrated continous-time filter design - an overview," *IEEE J. Solid-State Circuits*, vol. 29, no. 3, pp. 166–176, 1994.

[16] Rolf Schaumann, Mohammed S. Ghausi, and Kenneth R. Laker, *Design of Analog Filters, passive, active RC, and swithed capacitor*, Prentice-Hall, 1990.

[17] David A. Johns and Ken Martin, *Analog Integrated Circuit Design*, John Wiley & Sons, 1997.

[18] T.R. Viswanathan et al., "Switched-capacitor frequency control loop," *IEEE J. of Solid-State Circuits*, vol. 17, pp. 775–778, August 1982.

[19] University of Glasgow, *Xfilter User's Guide*, 1994.

Chapter 5

Precise Quadrature Signal Generation by Sampling

5.1 Introduction

As mentioned in Chapter two, digital wireless receivers require quadrature demodulation schemes to keep both amplitude and phase information of the received signal. Quadrature demodulation in digital domain was originally referred as quadrature sampling, a concept introduced by Grace and Pitt [1].

The conventional way of quadrature sampling is to use two analog mixers to generate a pair of quadrature signals at baseband, which are then sampled and digitised. This approach has been widely adopted in modern wireless communication systems due to its low requirement on image rejection filter.

The theorem of quadrature sampling of bandpass signals was established by Brown in [2]. Later on, various quadrature sampling techniques appeared [3, 4, 5, 6, 7, 8]. The basic idea for all these techniques is to sample and digitise signals at IF stage, then generate the quadrature signal components by different digital methods, for example, Hilbert transformation. They are therefore referred as direct sampling digital coherent detection techniques. These schemes demand an image rejection filter in RF stage with high quality factor, because the image interferer is very close to the desired signal.

The quadrature sampling process can also be realized by analog circuits [9, 10, 11, 12, 13, 14]. The major advantage of analog quadrature sampling is the reduced requirements on the analog-to-digital (A/D) conversion. However, it has the disadvantage of I/Q mismatch.

The market demands compact and low power wireless receivers. Complex-IF receivers exhibit these properties since they do not need image rejection filters in principle. To implement digital quadrature sampling schemes in this kind of receivers, complex IF signal can be digitised by either two real bandpass $\Delta\Sigma$ A/D converters or one complex bandpass $\Delta\Sigma$ A/D converter. However, due to mismatches between the two analog signal paths, the image rejection performance usually can not fulfil the stringent requirements in telecommunication standards like GSM.

Quadrature sampling for complex IF receivers can also be realized by an analog circuit [15], which is referred as *double quadrature sampling* (DQS). With a DQS circuit, two $\Delta\Sigma$ modulators with only lowpass quantisation noise shaping function can be used to digitise the complex IF signal. A great amount of chip area and power consumption can be saved then. Again, complex IF receivers employing DQS have also the problem of channel mismatches which limits their image rejection performance.

Circuit techniques for improving the image rejection performance of complex IF receivers with either digital or analog quadrature sampling, are developed in this chapter.

First, a new double quadrature sampling circuit, which is immune to channel mismatches and clock phase errors, is proposed. This circuit still has channel mismatches, but they only generate a *self-image*. The self-image origins from the desired signal itself, not from another radio channel. 40 dB of self-image rejection, which can be reasonably achieved with careful design, is sufficient in many applications with exceptions to be discussed in the next chapter.

Second, a complex notch rejection filter is proposed to be added at IF stage of the complex IF receivers employing both digital and analog quadrature sampling schemes, to suppress the image interferer. This notch filter is a first order finite impulse response (FIR) Hilbert transformer. Mismatch free switched-capacitor (SC) circuit realization of this filter is found. For receivers with analog quadrature sampling, this filter is incorporated to the DQS circuit. Since there are two image rejection processes, this DQS circuit together with the image filter is named as *double image rejection sampling* (DIRS).

In the next section, the operational principles of digital and analog quadrature sampling schemes are presented. Their circuit realizations, complexity and image rejection issues are discussed and compared. Then, quadrature sampling schemes for complex IF receivers are presented and new schemes with improved image rejection are proposed followed by verification of these schemes by circuit simulations. The last section summaries this chapter.

5.2 Quadrature Sampling of Real Signals

5.2.1 Conventional Approach

There are several established methods for quadrature sampling of a real signal. The oldest method uses two analog mixers and analog lowpass filters, and the A/D conversion is done in the baseband, as shown in Figure 5.1(a). This method sets low requirements on the A/D converter and high requirements on the phase adjustment between the sine and cosine wave. The spectra developed in this method are shown in Figure 5.2(a). The required minimum sampling frequency of the A/D converter is equal to the signal bandwidth B.

5.2.2 Digital Quadrature Sampling

Many digital quadrature sampling approaches appeared in the 80's of last century. These approaches sample and digitise the incoming signal at IF stage by one A/D converter, and then generate quadrature output components by different methods. Based on the quadrature generation methods, they can be categorised into two classes. The first class is by using digital Hilbert transformer, as shown in Figure 5.1(b). The Hilbert transformer can be in the form IIR [5] or FIR [16, 6], in the form of phase splitter [5] or the delay-transformer pair [16]. The detail description of Hilbert transformers has been presented in Chapter three. The spectra developed in this method, using the infinite impulse response Hilbert transformer proposed in [5], which has an "care-band" of one-fourth of the sampling frequency, are shown in Figure 5.2(b). In the last step of Figure 5.2(b), the signal is de-sampled by a factor of four to obtain the baseband output. Note that the required minimum sampling frequency of the A/D converter is four times of the signal bandwidth.

The second class of digital quadrature generation method is to use a quadrature mixer [4, 7, 8] as shown in Figure 5.1(c). In this method, the sampling frequency f_s of the A/D converter is usually set as four times the input center frequency so that the I and Q digital local oscillator signals are simply $\cos(n\pi/2)$ and $\sin(n\pi/2)$, $n = 0, 1, 2\ldots$ respectively. Since $\cos(n\pi/2) = 1, 0, -1, 0, \ldots$ and $\sin(n\pi/2) = 0, 1, 0, -1, \ldots$ for $n = 0, 1, 2\ldots$, the mixing process is carried out just by setting the even or odd samples to zero and altering the sign of the remaining samples. This process shifts the signal spectrum to baseband and $f_s/2$. Decimation by a factor of two is required to eliminate image at $f_s/2$ and to obtain I and Q data at a rate of $f_s/2$. If sub-sampling mode is used, then

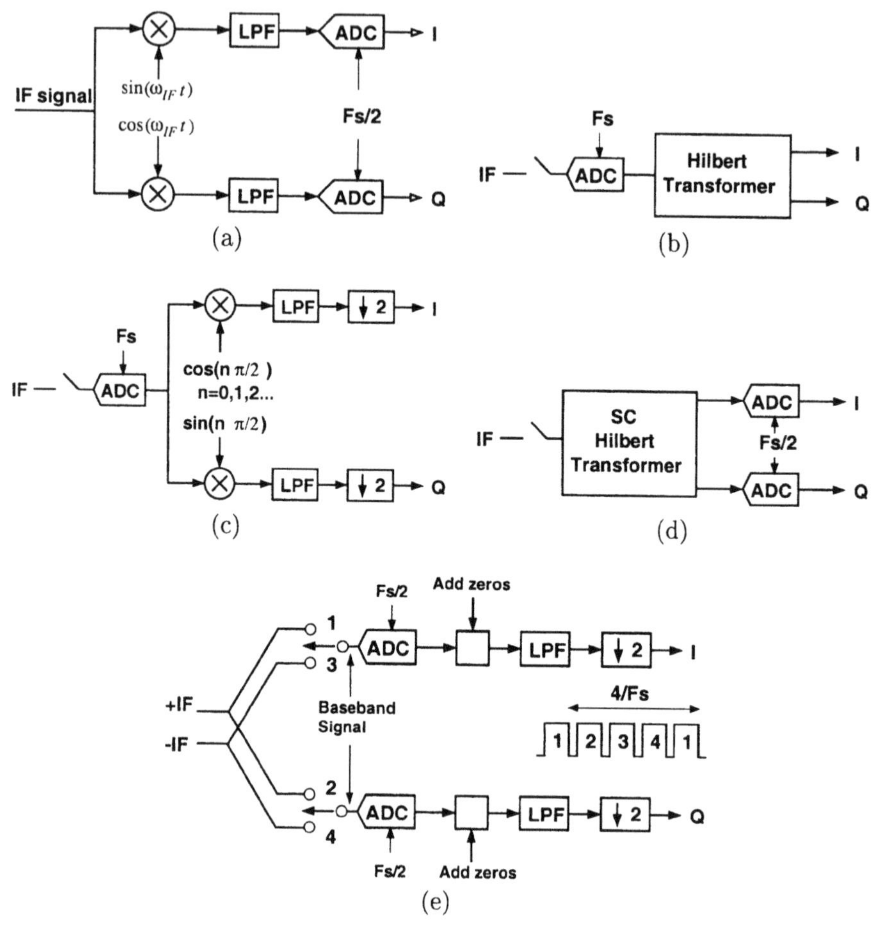

Figure 5.1: Different approaches of quadrature sampling. (a) the conventional approach; (b-c) digital quadrature sampling; (d-e) analog quadrature sampling.

the sampling frequency of A/D converter can be chosen as:

$$f_s = \frac{f_i}{N + \frac{2M+1}{4}} \tag{5.1}$$

where N and M are positive integers, f_i is the center frequency of the IF input signal. Obviously, the f_s must higher than the Nyquist rate, i.e., two times of the signal bandwidth B. The signal is shifted to $f_s/4$ after sub-sampling. The

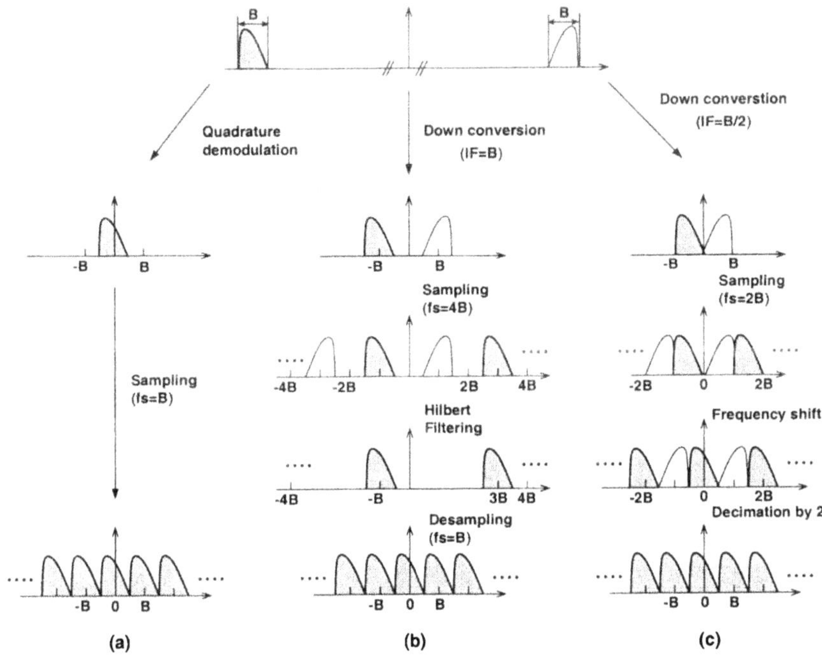

Figure 5.2: Spectra developed in different quadrature sampling approaches. (a-c) corresponds to the methods shown in Figure 5.1(a-c).

spectra developed in this method are shown in Figure 5.2(c). Note that the required minimum sampling frequency of the A/D converter is two times of the signal bandwidth.

The advantage of digital quadrature sampling methods over the convention method includes the perfect quadrature signal generation and the immunity to DC error, $1/f$ noise and other low frequency noises. However, they put higher demand on the speed of A/D converter and the image rejection filter. Moreover, in the case that oversampling $\Delta\Sigma$ A/D converter is used, it requires bandpass quantisation noise shaping function. The complexity of a bandpass $\Delta\Sigma$ A/D converter is two times that of a baseband $\Delta\Sigma$ A/D converter for the same degree of noise shaping [17]. By the way, it is worth to point out that almost all reported bandpass $\Delta\Sigma$ A/D converters are designed with sampling frequency being four times of the input center frequency in order to implement the quadrature sampling method of Figure 5.1(c) [18, 19, 20, 21, 22].

5.2.3 Analog Quadrature Sampling

The digital quadrature generation methods, both Figure 5.1(b) and (c), can be also realized by discrete-time circuit techniques.

In the method of Figure 5.1(b), if the Hilbert transformer is put before the A/D converter, we obtain an analog quadrature sampling scheme as shown in Figure 5.1(d). The Hilbert transformer can be realized by switched-capacitor techniques as described in Chapter three. In Figure 5.1(d), two A/D converters at lower sampling rate are required. Alternatively, one A/D converter at higher speed and with time-domain input multiplexing scheme can be employed to minimise the I/Q mismatch. An example of this quadrature sampling method can be found in [23].

In the method of Figure 5.1(c), the digital mixing process can be also realized in the analog domain by a special sampling technique as shown in Figure 5.1(e) [9, 10, 11, 12, 13, 14]. It is a multi-rate system. The even and odd samples are injected to I and Q channel respectively, and their sign are altered every other samples. After the A/D conversion, zeros are inserted between every two samples. Then the interpolated signals are low-passed, de-sampled by a factor of two to obtain the final outputs. Note that an inverting input is required in Figure 5.1(e). In differential circuits, this can be obtained easily by exchanging positive and negative input terminals.

The method shown in Figure 5.1(e) can be also viewed as that the incoming signal being sampled at IF stage by two sampling signals $p_I(t)$ and $p_Q(t)$ given by:

$$p_I(t) = \sum_{n=-\infty}^{+\infty} [\delta(t-nT) - \delta(t-nT-T/2)] \tag{5.2}$$

$$p_Q(t) = \sum_{n=-\infty}^{+\infty} [\delta(t-nT-T/4) - \delta(t-nT+T/4)] \tag{5.3}$$

where $T = 1/f_s$ and $\delta(t)$ is the Dirac function. The waveforms of $p_I(t)$ and $p_Q(t)$ are shown in Figure 5.3(a). Sampling by $p_I(t)$ and $p_Q(t)$ results in IF-to-baseband frequency translation, as explained in the following. Expanding $p_I(t)$ and $p_Q(t)$ into Fourier series, we have

$$p_I(t) = \sum_{k=-\infty}^{+\infty} a_k e^{jk\omega_0 t} dt, \quad p_Q(t) = \sum_{k=-\infty}^{+\infty} b_k e^{jk\omega_0 t} dt, \tag{5.4}$$

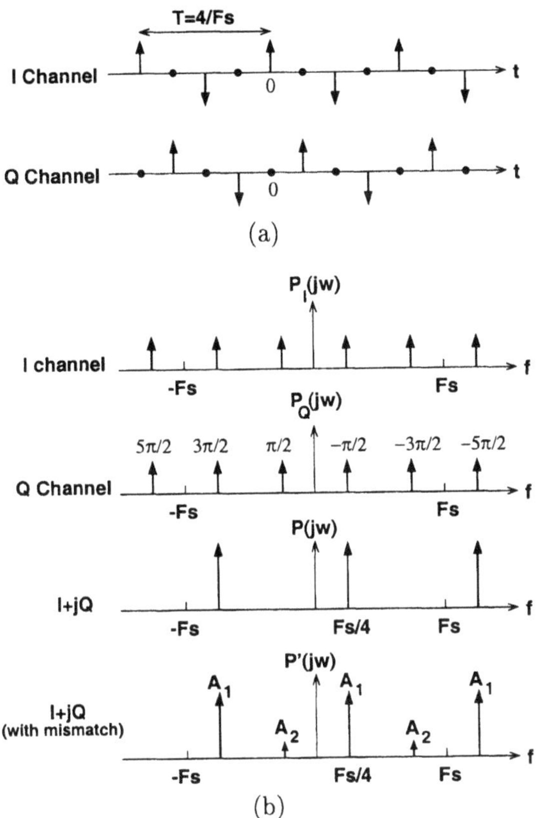

Figure 5.3: The I and Q sampling signals in the analog quadrature sampling scheme: (a) time waveforms; (b) their individual spectrum ($P_I(j\omega)$ and $P_Q(j\omega)$), and equivalent complex spectrum ($P_I(j\omega)+P_Q(j\omega)$) without and and with mismatches.

where $\omega_o = 2\pi/T = 2\pi f_s/4$, the coefficients a_k are given by

$$\begin{aligned} a_k &= \frac{1}{T}\int_{-T/2}^{T/2} p_I(t)e^{-jk\omega_0 t}dt \\ &= \frac{1}{T}\left(1 - e^{-jk\omega_0 T/2}\right) \\ &= \frac{1}{T}\left(1 - e^{-jk\pi}\right) \\ &= \begin{cases} \frac{2}{T} & \text{for } k = 2n+1,\ n = 0, \pm 1, \pm 2, \cdots \\ 0 & \text{otherwise,} \end{cases} \end{aligned} \tag{5.5}$$

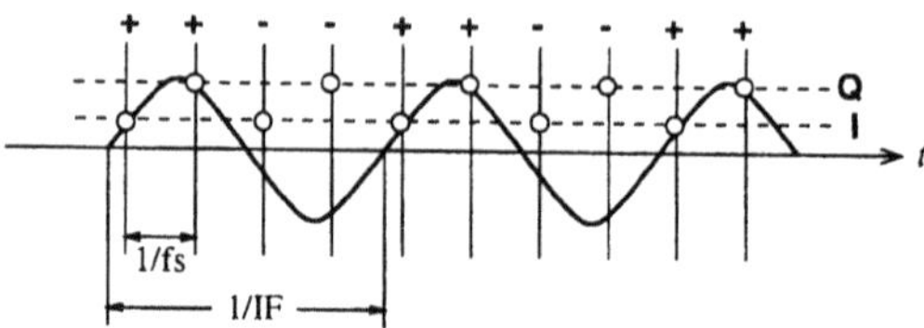

Figure 5.4: Quadrature sampling on a sinusoidal wave with a period of four times the sampling period.

and the coefficients b_k are given by

$$\begin{aligned} b_k &= \frac{1}{T}\int_{-T/2}^{T/2} p_Q(t)e^{-jk\omega_0 t}dt \\ &= \frac{1}{T}\left(e^{-jk\omega_0 T/4} - e^{+jk\omega_0 T/4}\right) \\ &= \frac{1}{T}\left(e^{-jk\pi/2} - e^{+jk\pi/2}\right) \\ &= \begin{cases} \frac{2}{T}e^{-jk\pi/2} & \text{for } k = 2n+1,\ n = 0, \pm1, \pm2, \cdots \\ 0 & \text{otherwise.} \end{cases} \end{aligned} \tag{5.6}$$

Therefore, the Fourier transforms $P_I(j\omega)$ and $P_Q(j\omega)$ of $p_I(t)$ and $p_Q(t)$ respectively, are obtained as:

$$P_I(j\omega) = \sum_{k=-\infty}^{+\infty} 2\pi a_k \delta(\omega - k\omega_0) \tag{5.7}$$

$$P_Q(j\omega) = \sum_{k=-\infty}^{+\infty} 2\pi b_k \delta(\omega - k\omega_0). \tag{5.8}$$

From (5.5) and (5.6), it can be observed that the the $p_I(t)$ and $p_Q(t)$ have nonzero frequency components at $(n \pm 1/4)f_s$, $n = 0, \pm1 \cdots$ only. The spectra $P_I(j\omega)$ and $P_Q(j\omega)$ are shown in Figure 5.3(b). By adding $jP_Q(j\omega)$ to $P_I(j\omega)$, we obtain the equivalent complex sampling spectrum as shown in Figure 5.3(b). Explicitly, the Fourier coefficients c_k of the complex sampling waveform $[p_I(t) + jp_Q(t)]$ are given by

$$\begin{aligned} c_k &= a_k + jb_k \\ &= \frac{1}{T}\left[1 - e^{-jk/\pi} + j\left(e^{-jk\pi/2} - e^{+jk\pi/2}\right)\right], \\ &= \begin{cases} \frac{4}{T}, & \text{for } k = 4n+1,\ n = 0, \pm1, \pm2, \cdots \\ 0, & \text{otherwise.} \end{cases} \end{aligned} \tag{5.9}$$

From the above equation, it can be found that $[p_I(t) + jp_Q(t)]$ has nonzero frequency components at $(n+1/4)f_s, n = 0, \pm 1 \cdots$ only. The resulted asymmetric frequency spectrum is due to the complex nature of this process. Sampling in time domain is equivalent to convolution in frequency domain. As illustrated in Figure 5.7(a)-(d), the IF spectrum is translated to baseband by convolving with the sampling spectrum.

The IF-to-baseband conversion of the sampling process can also be explained in the time-domain. Figure 5.4 illustrates the quadrature sampling process of a sinusoidal wave, $\cos(2\pi \frac{f_s}{4} t + \phi)$, where ϕ is an arbitrary initial phase and the signal frequency is one-fourth of the f_s. Odd and even samples are for I and Q channel respectively. At sampling instance ($t = n/f_s$, $n = 0, 1, 2 \ldots$), we have:

$$\text{I:} \qquad \cos(\frac{\pi f_s}{2} \frac{2n+1}{f_s} + \phi) = -\sin\phi, +\sin\phi, -\sin\phi, +\sin\phi, \ldots$$

$$\text{Q:} \qquad \cos(\frac{\pi f_s}{2} \frac{2n}{f_s} + \phi) = +\cos\phi, -\cos\phi, +\cos\phi, -\cos\phi \ldots$$

After sign alternation, the I and Q channel are equal to $-\sin\phi$ and $+\cos\phi$ respectively, for all n. So they are shifted to baseband.

Besides, the low frequency noise from the input is translated to $f_s/2$. Therefore, the analog quadrature sampling methods are immune to low-frequency noise, too. However, they are sensitivity to the $1/f$ noise of the opamp of the A/D converter, because this kind of noise is produced after the frequency translation.

In the analog quadrature sampling scheme shown in Figure 5.1(e), the required minimum sampling frequency of A/D converter is equal to signal bandwidth B. Remember that the minimum sampling frequency required in its digital counterpart shown in Figure 5.1(c) is $2B$. However this is not the advantage of Figure 5.1(e) because it requires two A/D converters. The great advantage of this method is that the signals at the input of A/D converters lie at baseband but not IF. So when $\Delta\Sigma$ A/D converters are employed, only low-pass quantisation noise shaping function is required. As mentioned earlier, the order of a baseband $\Delta\Sigma$ A/D converter is only the half of that of a bandpass $\Delta\Sigma$ A/D converter of the same degree of noise shaping. The main disadvantage of this method is that it involves two analog paths and therefore the I/Q mismatch problem.

A switched-capacitor realization of this quadrature sampling scheme is shown in Figure 5.5. Note that it is efficiently realized by the sampling unit of the baseband $\Delta\Sigma$ A/D converters. A 200MHz CMOS I/Q down-converter employing this scheme was reported with I/Q balance of better than 0.33 dB and 0.7^o [12].

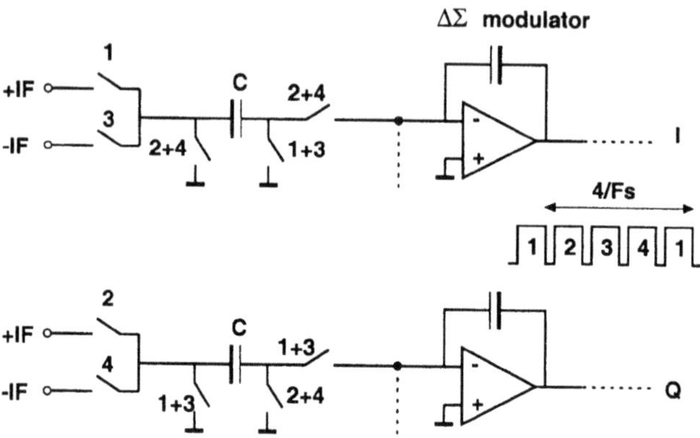

Figure 5.5: Quadrature sampling realized in the sampling units of two baseband $\Delta\Sigma$ A/D converters.

Mismatch Effects

In analog quadrature sampling circuits, mismatches between sampling capacitors and the gain difference between I and Q A/D converters are unavoidable. Such a mismatch is modelled in Figure 5.6, where G_i and G_q represents the gains of I and Q channels respectively. In general, $G_i \neq G_q$. $X(j\omega)$ is the input signal spectrum. Under this model, the output spectrum $Y(j\omega)$ can be written as:

$$\begin{aligned} Y(j\omega) &= G_i P_I(j\omega) \otimes X(j\omega) + jG_q P_Q(j\omega) \otimes X(j\omega) \\ &= P'(j\omega) \otimes X(j\omega), \end{aligned} \tag{5.10}$$

where

$$P'(j\omega) = \frac{G_i + G_q}{2} P(j\omega) + \frac{G_i - G_q}{2} P_I^*(j\omega) \tag{5.11}$$

and $P(j\omega) = P_I(j\omega) + jP_Q(j\omega)$, $\otimes$ represents convolution and superscript * represents complex conjugate.

Equation (5.11) tells us that with I/Q mismatch, the equivalent sampling waveform $P'(j\omega)$ contains a normal term $\frac{G_i+G_q}{2}P(j\omega)$ and a mirror term $\frac{G_i-G_q}{2}P_I^*(j\omega)$. The spectrum of $P'(j\omega)$ is shown in Figure 5.3(b), where frequency components at $(n+\frac{1}{4})f_s, n = 0, \pm 1, \ldots$ represent the mirror term $\frac{G_i-G_q}{2}P_I^*(j\omega)$.

We denote the magnitude of the desired frequency components of the sampling waveform at $(n+\frac{1}{4})f_s, n = 0, \pm 1 \ldots$ as A_1, and those undesired at

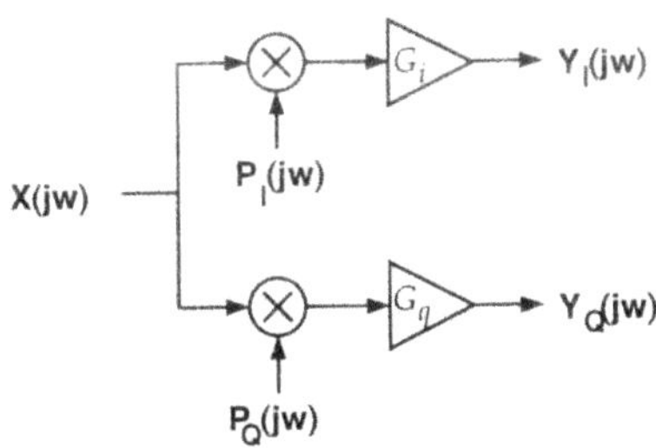

Figure 5.6: Mismatch model of the quadrature sampling circuit.

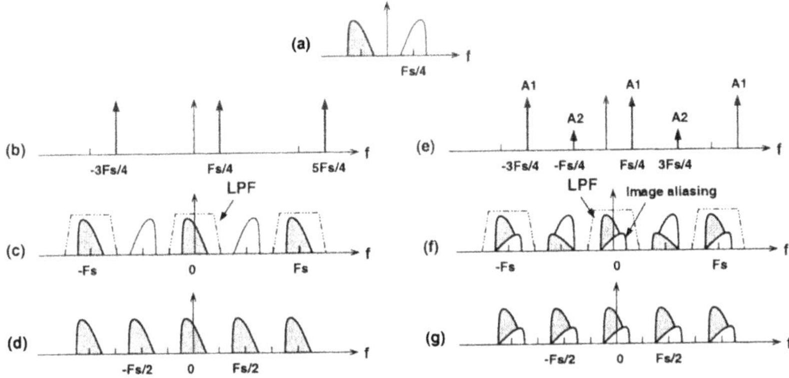

Figure 5.7: Spectra developed in analog quadrature sampling process: (a) IF input signal; (b) sampling signal; (c) the result of mixing between (a) and (b); (d) after decimation by a factor of 2. (e-g) same process as (b-d) but with I/Q mismatches present.

$n - f_s/4, n = 0, \pm 1 \ldots$ as A_2. From (5.11) we have:

$$\frac{A_1}{A_2} = \frac{G_i - G_q}{G_i + G_q}. \tag{5.12}$$

Sampling by this waveform will cause the image of the signal to be folded to the desired signal band, as illustrated in Figure 5.7(e)-(h). Since this image is the mirrored version of the desired signal itself, the mismatch effect is not very important. Only signal-to-noise ratio (SNR) of the receiver will be degraded. Let G_i, $G_q = G \pm \Delta G$, where G is the nominal gain value and ΔG is the deviation. The self-image rejection ratio (IRR) is equal to A_2^2/A_1^2 in this system.

From (5.12), we have

$$IRR = A_2^2/A_1^2 \cong (\frac{2G}{2\Delta G})^2 = \frac{1}{(\Delta G/G)^2}. \tag{5.13}$$

For example, for 1% mismatch, i.e., $\Delta G/G = 0.01$, the self-image rejection ratio is limited to 40 dB.

The above analysis assumes fixed mismatch. Alternatively, it can also be analysed by statistical method. Assume G_i and G_q are independent Gaussian distributed random variables with the same mean and variance. If we normalise the mean to one and denote the adjusted variance as σ_0^2, then σ_0^2 represents the mismatch between G_i and G_q. Then,

$$\begin{aligned} IRR &= \overline{A_2^2/A_1^2} = \overline{(G_i^2 + G_q^2 + 2G_iG_q)}/\overline{(G_i^2 + G_q^2 - 2G_iG_q)} \\ &= \frac{4 + 2\sigma_0^2}{2\sigma_0^2} \cong \frac{2}{\sigma_0^2}, \end{aligned} \tag{5.14}$$

where $\overline{\bullet}$ represents statistical average.

Phase Error

In addition to the mismatch problem, phase errors can also present if the sampling instant for I and Q channels are not apart from each other by exactly half a clock cycle. This error is referred to as *non-even phase error.* Assuming the timing error be τ, $\tau \ll T/4$, then the sampling waveform $p_Q(t)$ of equation (5.3) becomes:

$$p'_Q(t) = \sum_{n=-\infty}^{+\infty} [\delta(t - nT - T/4 - \tau) - \delta(t - nT + T/4 - \tau)]. \tag{5.15}$$

The Fourier series coefficients b'_k of $p'_Q(t)$ become:

$$\begin{aligned} b'_k &= \frac{1}{T}\int_{-T/2}^{T/2} p'_Q(t)e^{-jk\omega_0 t}dt \\ &= \frac{1}{T}\left(e^{-jk\omega_0(T/4+\tau)} - e^{+jk\omega_0(T/4-\tau)}\right) \\ &= \frac{e^{-j2\pi k\tau/T}}{T}\left(e^{-jk\pi/2} - e^{+jk\pi/2}\right). \end{aligned} \tag{5.16}$$

Defining $\theta = 2\pi\tau/T$ as the phase error, the Fourier series coefficients c'_k of $[p_I(t) + jp'_Q(t)]$ become:

$$c'_k = a_k + jb'_k = \frac{1}{T}\left[1 - e^{-jk/\pi} + je^{-jk\theta}\left(e^{-jk\pi/2} - e^{+jk\pi/2}\right)\right]. \tag{5.17}$$

From the above equation, it can be found that frequency components of the complex sampling signal $p_I(t) + jp'_Q(t)$ at $(n - \frac{1}{4})f_s$, $n = 0, \pm1, \cdots$ are not exactly cancelled out. We have $c_1 = \frac{2}{T}(1 + e^{-j\theta})$ and $c_{-1} = \frac{2}{T}(1 - e^{-j\theta})$. The power ratio between c_1 and c_{-1} is the self-image rejection ratio:

$$\mathrm{IRR} = \frac{|c_1|^2}{|c_{-1}|^2} \cong \frac{1}{(\theta/2)^2}. \tag{5.18}$$

For example, 1° phase error gives a self-image rejection ratio of 41 dB. It is worthy to point out that if the circuit is operated in sub-sampling mode, then the effect of phase error will get worse. In such case, the IRR, given by the power ratio of relevant coefficients c_k, is getting worse with increasing k. This effect can be understood intuitively. For higher frequency input, the same amount of timing error gives a greater sampling error.

The non-even phase error can be avoided by using only one sample-and-hold (S/H) circuit clocking at f_s, and passing even and odd samples to I and Q channels respectively. Figure 5.8 shows such a S/H circuit [24].

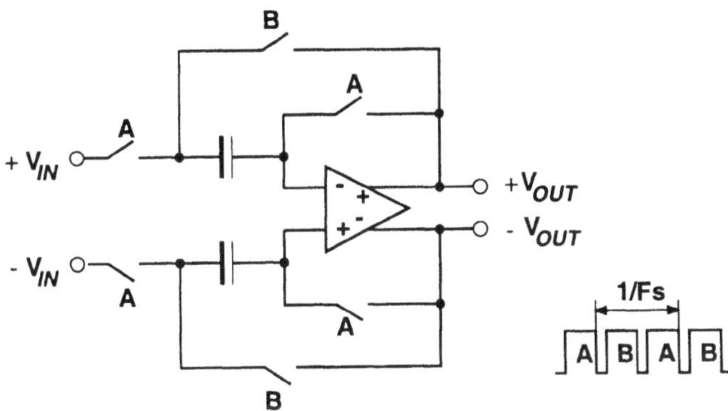

Figure 5.8: A mismatch-free sample-and-hold circuit.

5.3 Quadrature Sampling of Complex Signals

The market demands compact and low power wireless receivers. The complex-IF architecture [25], also known as Weaver's image rejection architecture [26], exhibits these properties since it does not need image rejection filters in principle. The block diagram of this receiver is shown in Figure 5.9. The image cancellation principle of this receiver has been presented in Chapter two.

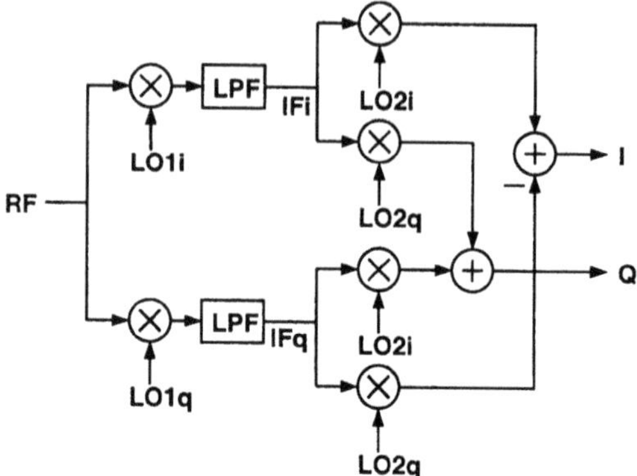

Figure 5.9: Architecture of the complex-IF receiver.

The sampling of complex IF signals in quadrature form is called as *double quadrature sampling* in this book. As for quadrature sampling, there are also both analog and digital approaches for DQS. The digital approach requires digitisation directly at IF stage. The analog approach, however, samples at IF stage but digitises the signal at baseband. These DQS schemes are discussed in the following subsections.

5.3.1 IF-Sampling IF-Digitising

One DQS method is to sample and digitise the complex input signals at IF stage, and then perform the complex IF-to-baseband conversion in digital domain. This method is also referred as direct IF digitising. Figure 5.10(a-b) shows two variations of this method. Figure 5.10(a) uses two A/D converters with sampling frequency at least two times the input signal bandwidth. Since the signals at the input of A/D converters are centered at one fourth of the sampling frequency, bandpass quantisation noise shaping function must be employed if oversampled A/D converters are used.

In Figure 5.10(b), a complex bandpass $\Delta\Sigma$ modulator for digitising the complex IF signal is used instead. The complex bandpass $\Delta\Sigma$ modulator is so called because their noise and signal transfer functions are complex, asymmetric [27, 28, 29]. Notches of their noise transfer function appears mainly in one side of the frequency axis. Therefore, it reduces the order of the noise transfer function and therefore the circuit complexity.

To understand the complex IF-to-baseband conversion shown in Figure 5.10(a-b), we denote the complex input as $X_I + jX_Q$, and the complex LO signal as

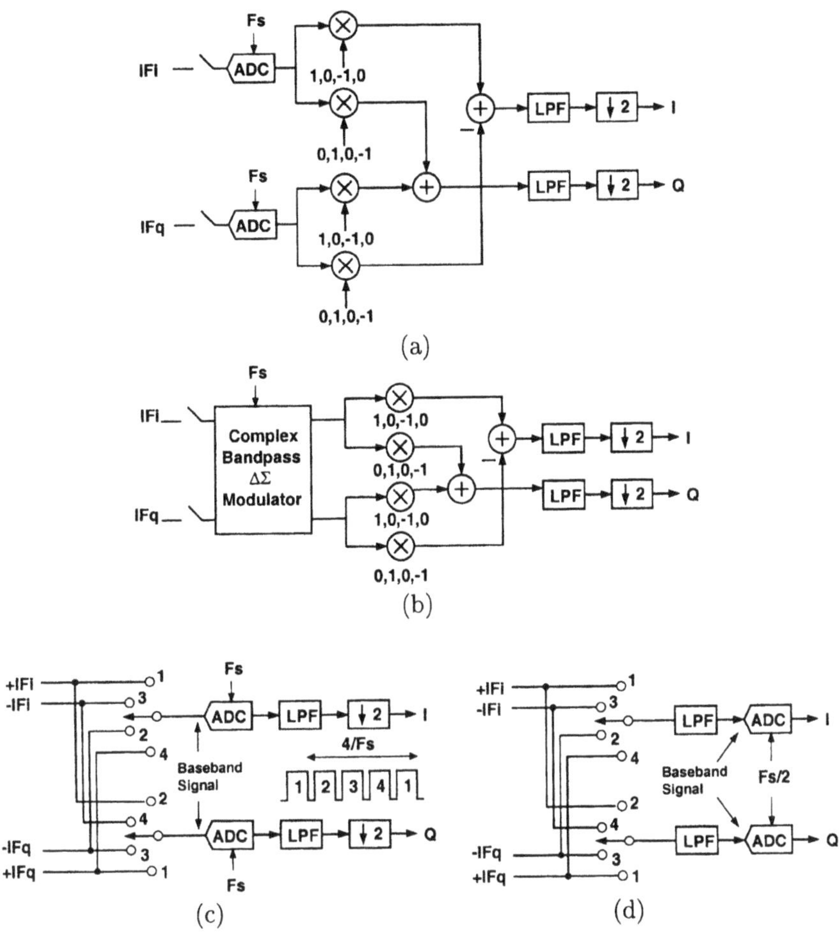

Figure 5.10: Double quadrature sampling approaches (a-b) in digital domain; (c-d) in analog domain.

$\cos(n\pi/2) + j\sin(n\pi/2)$, where $n = 0, 1, 2 \ldots$. Multiplying the input with the LO, we have the output as

$$\begin{aligned} & (X_I + jX_Q)[\cos(n\pi/2) + j\sin(n\pi/2)] \\ = \; & [X_I\cos(n\pi/2) - X_Q\sin(n\pi/2)] + j[X_Q\cos(n\pi/2) + X_I\sin(n\pi/2)] \end{aligned} \tag{5.19}$$

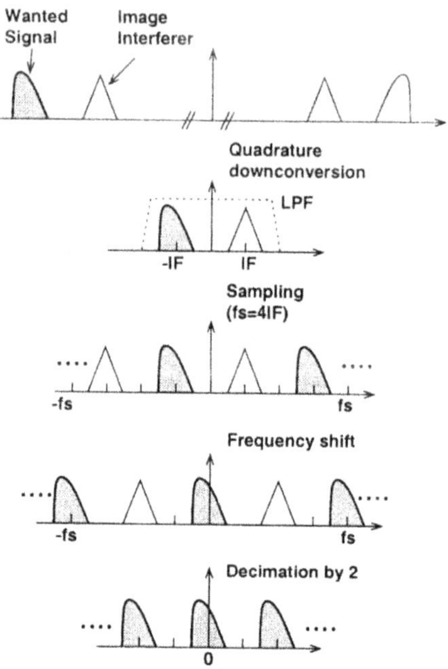

Figure 5.11: Spectra developed in double quadrature sampling.

The first and second terms of the above equation are the I and Q components of the output. Therefore, we obtain the IF-to-baseband conversion configuration shown in Figure 5.10(a-b).

The spectra developed in these methods are shown in Figure 5.11. As illustrated in the figure, high-Q and higher frequency bandpass filter is no longer required for suppressing the image signals. Instead, only a lowpass anti-aliasing filter is required before the sampling stage. An interferer at the image frequency is translated to $f_s/2$ and filtered in digital domain.

Mismatch Effects

The DQS methods shown in Figure 5.10 (a-b) have two parallel analog paths and therefore have the the problem of channel mismatch.

Let us consider the Figure 5.10(a) first. The mismatch lies on the gain imbalance between two A/D converters. This mismatch distorts the IF input signal. To see the effect, let us denote the I and Q input spectrum as $X_I(j\omega)$ and $X_Q(j\omega)$ respectively, and let the gain of I and Q A/D converters be G_i

and G_q respectively. With the mismatch, the input spectrum becomes

$$\begin{aligned} & G_i X_I(j\omega) + jG_q X_Q(j\omega) \\ = \; & \frac{G_i + G_q}{2}[X_I(j\omega) + jX_Q(j\omega)] - \frac{G_i - G_q}{2}[X_I(j\omega) - jX_Q(j\omega)] \\ = \; & \frac{G_i + G_q}{2} X(j\omega) - \frac{G_i - G_q}{2} X^*(j\omega) \end{aligned} \tag{5.20}$$

where $X(j\omega) = X_I(j\omega) + jX_Q(j\omega)$. The first term of (5.20) represents the nominal input and the second term of (5.20) represents a mirrored version of first one. Therefore, a portion of the image interferer is folded to the wanted signal band, and vice versa. We define the image rejection ratio (IRR) as the ratio of the power gain of the desired signal $X(j\omega)$ over the power gain of the image signal $X^*(j\omega)$. From (5.20), we have

$$IRR = \frac{(G_i + G_q)^2}{(G_i - G_q)^2}. \tag{5.21}$$

Assume G_i and G_q are independent Gaussian distributed random variables with the same mean and variance. As before, we normalise the mean to one and denote the adjusted variance as σ_0^2 which represents the mismatch between G_i and G_q. Then we have

$$IRR = \overline{(G_i + G_q)^2} / \overline{(G_i - G_q)^2} = \frac{4 + 2\sigma_0^2}{2\sigma_0^2} \cong \frac{2}{\sigma_0^2}. \tag{5.22}$$

For the DQS method shown in Figure 5.10(b), the mismatch effect is same as that given in (5.22) [30]. Lastly, both Figure 5.10(a) and (b) do not have the problem of non-even phase error since the input is sampled in one clock phase only.

5.3.2 IF-Sampling Baseband-Digitising

Another DQS method is to sample the complex input signals at IF stage, perform the complex IF-to-baseband conversion by a special sampling technique, then the A/D conversions in baseband, as shown in Figure 5.10(c). This method is obtained actually by moving forward the complex IF-to-baseband conversion before the A/D conversion. Since signals at the input of A/D converters lie in baseband, lowpass quantisation noise shaping function can be adopted in case of oversampling $\Delta\Sigma$ A/D conversion.

In Figure 5.10(c), IF inputs $+IF_i$,$+IF_q$,$-IF_i$, and $-IF_q$ are sampled by I channel during clock phases 1, 2, 3 and 4 respectively; $+IF_q$,$+IF_i$,$-IF_q$, and $-IF_i$ are sampled by Q channel during clock phases 1, 2, 3 and 4 respectively. We can find that the I and Q outputs are equivalent to $[X_I \cos(n\pi/2) - X_Q \sin(n\pi/2)]$ and $[X_Q \cos(n\pi/2) + X_I \sin(n\pi/2)]$, $n = 0, 1, 2, \ldots$ shown in (5.19). The spectra developed in this method is same as that shown in Figure 5.11. The difference from the previous method is that the frequency translation is carried out in the discrete-time analog instead of the digital domain.

A circuit realization of the special sampling technique is shown in Figure 5.12 [15]. We see that it is efficiently realized by the sampling units of $\Delta\Sigma$ A/D converters. Moreover, the function of sub-sampling can be also realized by this circuit if desired.

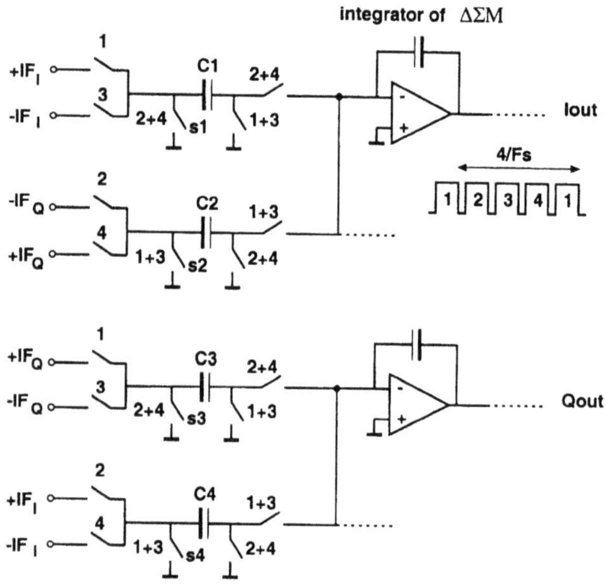

Figure 5.12: Double quadrature sampling realized in the sampling units of two baseband $\Delta\Sigma$ A/D converters.

Figure 5.10(d) is an variation of Figure 5.10(c). It moves the the 2-fold decimator further to the analog portion of the system, so that the sampling rate of the A/D converters can be reduced by a factor of two. The lowpass filter, or more precisely, the half-band filter, can be implemented by switched-capacitor circuits. The disadvantage of this scheme is that implementing the filters in analog domain may introduce more I/Q mismatches. But for large oversampling ratio ($B \ll f_s$), a very simple notch filter, with transfer function of $H_1(z) = 1+z^{-1}$, can be used. This notch filter can be realized just by adding odd samples to even samples. Obviously, this analog decimation method will not introduce more mismatches.

Mismatch Effects

The DQS methods shown in Figure 5.10(c-d) have four analog paths. To analyse the mismatch effect, let us model the sampling process as shown in Figure 5.13, where the P_I and P_Q are the I and Q sampling signals shown in Figure 5.3. Denoting the gain in four paths as G_{ii}, G_{iq}, G_{qi} and G_{qq} as shown in Figure 5.13, the output $Y(j\omega)$ is then given by

$$\begin{aligned} Y(j\omega) &= Y_I(j\omega) + jY_Q(j\omega) \\ &= G_{ii}P_I(j\omega) \otimes X_I(j\omega) - G_{qq}P_Q(j\omega)X_Q(j\omega) \\ &\quad +j\left(G_{iq}P_Q(j\omega) \otimes X_I(j\omega) + G_{qi}P_I(j\omega) \otimes X_Q(j\omega)\right) \\ &= P_{cm}(j\omega) \otimes X(j\omega) + P_{diff}(j\omega) \otimes X^*(j\omega), \end{aligned} \tag{5.23}$$

where

$$P_{cm}(j\omega) = \frac{G_{ii} + G_{qi} + G_{iq} + G_{qq}}{4} P(j\omega) + \frac{G_{ii} + G_{qi} - (G_{iq} + G_{qq})}{4} P^*(j\omega) \tag{5.24}$$

$$P_{diff}(j\omega) = \frac{G_{ii} - G_{qi} + G_{iq} - G_{qq}}{4} P(j\omega) + \frac{G_{ii} - G_{qi} - (G_{iq} - G_{qq})}{4} P^*(j\omega). \tag{5.25}$$

and $X(j\omega) = X_I(j\omega) + jX_Q(j\omega)$, $P(j\omega) = P_I(j\omega) + jP_Q(j\omega)$. The output consists of two parts: input $X(j\omega)$ sampled by P_{cm} and input conjugate $X^*(j\omega)$ sampled by P_{diff}. The spectra of P_{cm}, P_{diff} and output $Y(j\omega)$ are shown in Figure 5.14.

From equation (5.23), it can be found that the baseband output consists of a wanted signal and three unwanted interferers, namely, $n1$, $n2$ and $n3$ as shown in Figure 5.14. Among these interferers, n_3 is the image of the wanted signal itself, it affects only the signal to noise ratio and is not harmful except in multi-band receivers. Moreover, this kind of interferer can be corrected by digital correction methods, which will be addressed in the next chapter. The problematic interferers are n_1 and n_2. They origin from another radio channel that could be 80 dB stronger than the desired channel as allowed in some wireless standards such as GSM. We define the image rejection ratio as the ratio of the power gain of the desired signal over the power gain of images n_1

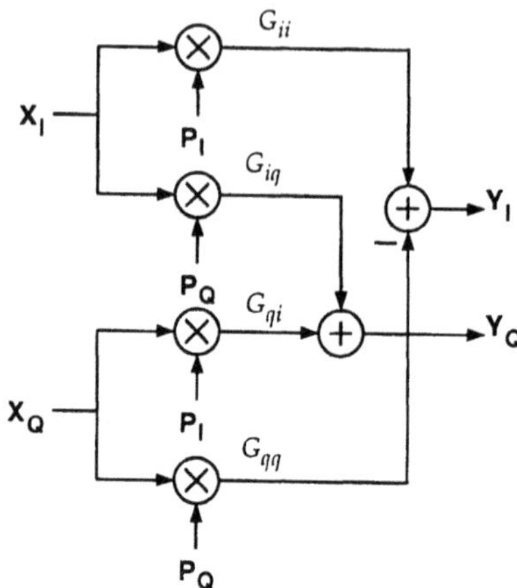

Figure 5.13: The mismatch model of the DQS method of Figure 5.10(c).

and n_2. From equation (5.23), we have

$$\begin{aligned} IRR &= \frac{(G_{ii}+G_{qi}+G_{iq}+G_{qq})^2}{(G_{ii}-G_{qi}+G_{iq}-G_{qq})^2+[G_{ii}-G_{qi}-(G_{iq}-G_{qq})]^2} \\ &= \frac{(G_{ii}+G_{qi}+G_{iq}+G_{qq})^2}{2(G_{ii}-G_{qi})^2+2(G_{iq}-G_{qq})^2}. \end{aligned} \tag{5.26}$$

Assume G_{ii}, G_{qi}, G_{iq}, and G_{qq} are independent Gaussian distributed random variables with the same mean and variance. As before, we normalise the mean to one and denote the adjusted variance as σ_0^2 which represents the mismatch among G_{ii}, G_{qi}, G_{iq}, and G_{qq}. Then we have

$$IRR = \frac{\overline{(G_{ii}+G_{qi}+G_{iq}+G_{qq})^2}}{\overline{2(G_{ii}-G_{qi})^2+2(G_{iq}-G_{qq})^2}} = \frac{16+4\sigma_0^2}{2(2\sigma_0^2)+2(2\sigma_0^2)} \cong \frac{2}{\sigma_0^2}. \tag{5.27}$$

For example, if $\sigma_0^2 = 0.01$, we have 46 dB of image rejection. This might not be sufficient in many applications.

It is interesting to know, from equation (5.22) and (5.27), that both the IF-sampling IF-digitising and IF-sampling baseband-digitising DQS methods have the same image rejection ratio under the same mismatch conditions, although the latter has more analog paths.

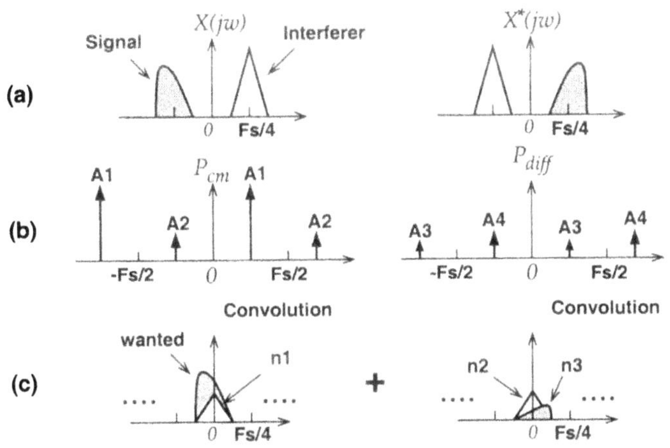

Figure 5.14: Spectra developed in analog double quadrature sampling scheme. (a) input and its complex conjugate; (b) common-mode and differential sampling signal, where $A_1 = \frac{G_{ii}+G_{qi}+G_{iq}+G_{qq}}{4}$, $A_2 = \frac{G_{ii}+G_{qi}-(G_{iq}+G_{qq})}{4}$, $A_3 = \frac{G_{ii}-G_{qi}+G_{iq}-G_{qq}}{4}$ and $A_4 = \frac{G_{ii}-G_{qi}-(G_{iq}-G_{qq})}{4}$; and (c) outputs.

Non-even Phase Error

Lastly, non-even phase error can also lead to image problem in the DQS methods shown in Figure 5.10(c-d). The image rejection ratio as a function of phase error is the same as that given in (5.18). Again, this problem can be avoided by employing in I and Q channels a sample-and-hold stage as shown in Figure 5.8, which is clocked at f_s. Note that the circuit of Figure 5.8 is free of capacitance mismatch, therefore adding this circuit to I and Q channels will introduce no mismatches.

5.4 Image Rejection Improvement Methods

Two new circuit techniques that improve the image rejection performance of double quadrature sampling scheme are presented as follows.

5.4.1 A Mismatch and Phase Error Free DQS Circuit

A DQS circuit for IF-sampling baseband-digitising, immune to channel mismatch and phase errors, was found [31]. The proposed circuit is shown in Figure 5.15, which differs only slightly from the one shown in Figure 5.12. How it avoids the problems of channel mismatches and phase errors is explained as follows.

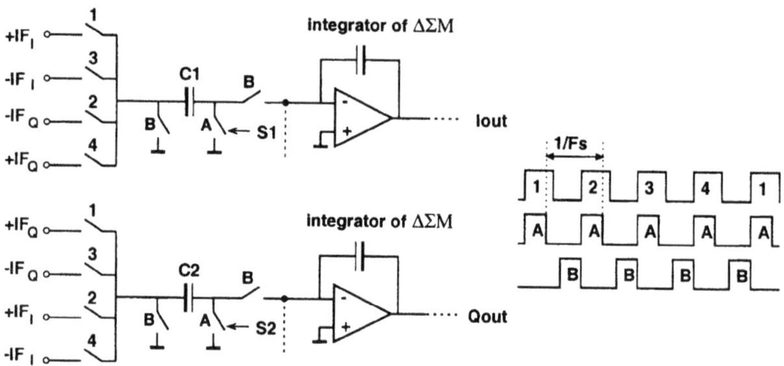

Figure 5.15: A mismatch and phase error free DQS circuit and the phase diagram.

First, it reduces the number of sampling paths from four in Figure 5.12 to two here. This is accomplished by halving the pulse width of the clock phase 1, 2, 3, and 4, and by using additional clock phase A and B with frequency of f_s as shown in the phase diagram of Figure 5.15. The capacitors C_1 and C_2 are shared in all the four sampling phases for channel I and Q respectively. Mismatch exists between C_1 and C_2. However, as to be explained later in this subsection, this mismatch produces only self-image which origins not from another radio channel but from the desired signal itself. 40 dB of image rejection can be very sufficient in this case.

Second, the effect of phase errors is avoided by making switches S_1 and S_2, controlled by clock A, opened a little earlier than those switches connecting to inputs. This arrangement is usually adopted in switched-capacitor circuits for the removal of the clock feed-through and charge injection effects of sampling switches [32]. More importantly here, this arrangement makes the sampling instant being determined only by the falling edge of phase A. When switch S_1 or S_2 opens, then capacitor C_1 or C_2 becomes floating and stops charging to the inputs, even when the input switches are still on. Therefore, the effect of relative timing errors among clock phase 1, 2, 3 and 4 is avoided.

To analyse the effect of mismatch between C_1 and C_2, let us refer to the model shown in Figure 5.13. Since the same capacitor C_1 or C_2 is shared in all the sampling phases for I and Q outputs respectively, we have $G_{ii} = G_{qq}$ and

$G_{iq} = G_{qi}$. So the P_{cm} and P_{diff} shown in (5.24) and (5.25) becomes:

$$P_{cm}(j\omega) = \frac{G_{ii} + G_{qi}}{2} P(j\omega) \quad (5.28)$$

$$P_{diff}(j\omega) = \frac{G_{ii} - G_{qi}}{2} P^*(j\omega). \quad (5.29)$$

Referring to Figure 5.14, A_2 and A_3 become zero. The image terms $n1$ and $n2$ are eliminated. Only the self-image component n_3 remains. The self-image problem is not a serious problem except in receivers that convert multi-channel signals to digital domain. This issue will be discussed in the next two chapters and solutions will be provided.

Total image rejection can be achieved in this IF-to-baseband stage for complex IF receivers. Only the less-important self-image problem arises from channel mismatches.

Second Order Effects

Besides capacitor mismatch and clock phase error, second order circuit nonidealities that may affect the image rejection of the proposed circuit are discussed in the following paragraphs.

Let us consider the charge injection firstly. As they switch off a little bit later than switches S_1 and S_2 that are controlled by phase A, the input switches, controlled by either phase 1, or 2, or 3, or 4, do not produce charge injection across capacitor C_1 or C_2. Only those switches controlled by phase A or B do. However, since these switches are connected either to ground or virtual ground, their charge injections are signal-independent, constant for all the four sampling phases. Therefore, the charge injection will not affect the image rejection of the sampling circuit. Furthermore, this constant charge injection can be cancelled by fully differential circuit architecture.

Now consider the effect of the ON-resistance mismatch of sampling switches. This mismatch will lead to different *settling time* in the four sampling phases. Assume that the ON-resistances of all sampling switches are constant. A relative error in the ON-resistance, denoted as $\Delta R/R$, will give a relative error in the voltage charged to the sampling capacitor (C_1 or C_2), denoted as $\Delta V_c/V_c$. It can be easily proven that this error is given by the following formula:

$$\frac{\Delta V_c}{V_c} \cong e^{-\frac{T}{2RC}}\left(1 - e^{\frac{T}{2RC}\frac{\Delta R}{R}}\right), \quad \text{for } \frac{T}{2RC} \gg 1 \text{ and } \frac{\Delta R}{R} \ll 1, \quad (5.30)$$

where $T = 1/f_s$, R is the nominal ON-resistance, C is the capacitance of the sampling capacitor. The effect of $\Delta V_c/V_c$ is similar to the effect of channel mismatch. Substituting $\Delta V_c/V_c$ for $\Delta G/G$ in (5.27), we obtain the IRR with respect to the mismatch of switch ON-resistance. From (5.30), to reduce $\Delta V_c/V_c$, one must decrease the switch ON-resistance. Table 5.1 lists the image rejection ratio of the proposed circuit under 1% mismatch of switch ON-resistance and for different values of RC time constant. From the table, one can find that when the RC time constant is less than or equal to one-seventh of a half sampling period ($\frac{T}{2RC} \geq 7$), the IRR is larger than 80 dB and the ON-resistance mismatch effect can be ignored.

Table 5.1: IRR of the proposed circuit under 1% mismatch of switch ON-resistance.

	$\frac{T}{2RC} = 3$	$\frac{T}{2RC} = 5$	$\frac{T}{2RC} = 7$	$\frac{T}{2RC} = 10$
$\frac{\Delta V_c}{V_c}$	-0.152%	-0.0345%	-0.00661%	-0.000477%
IRR	56 dB	69 dB	84 dB	106 dB

Lastly, other kinds of circuit non-idealities, like different kT/C noise in I and Q paths, finite gain and bandwidth of amplifiers, are signal-independent and equal in all the four sampling phases. They will not affect the image rejection performance of the proposed circuit.

Simulation Results

To verify the proposed DQS circuit, circuit simulations were conducted by using SWITCAP2 [33].

Firstly, a third order cascaded $\Delta\Sigma$ modulator [34] with lowpass noise shaping is used here for performing the A/D conversion. The system diagram of the modulator is shown in Figure 5.16(a). It is a cascade of first-order modulators. The digitisers and D/A converters of the modulator have all 1-bit resolution. The switched-capacitor circuit of the first modulator is shown in Figure 5.16(b). Note that this modulator is selected arbitrarily for demonstration of the DQS schemes. Any other baseband modulators are applicable here as long as they can be realized by SC circuits.

To perform quadrature sampling of the IF signal, the sampling unit shown in Figure 5.16(b) is replaced by the proposed DQS circuit shown in Figure 5.15 or the old DQS shown in Figure 5.12 for comparison.

In the simulations, two complex tones with spectrum shown in Figure 5.17(a) are applied as input. The tone with frequency at $-(0.25+2^{-5})f_s$ is assumed to be the desired signal, and the other with frequency at $(0.25+2^{-6})f_s$ is assumed to be the image interferer. The responses of the old and new DQS schemes to this input are shown in Figure 5.17(b) and (c) respectively. Fixed capacitor mismatch of 1% and fixed clock phase error of 1° are assumed.

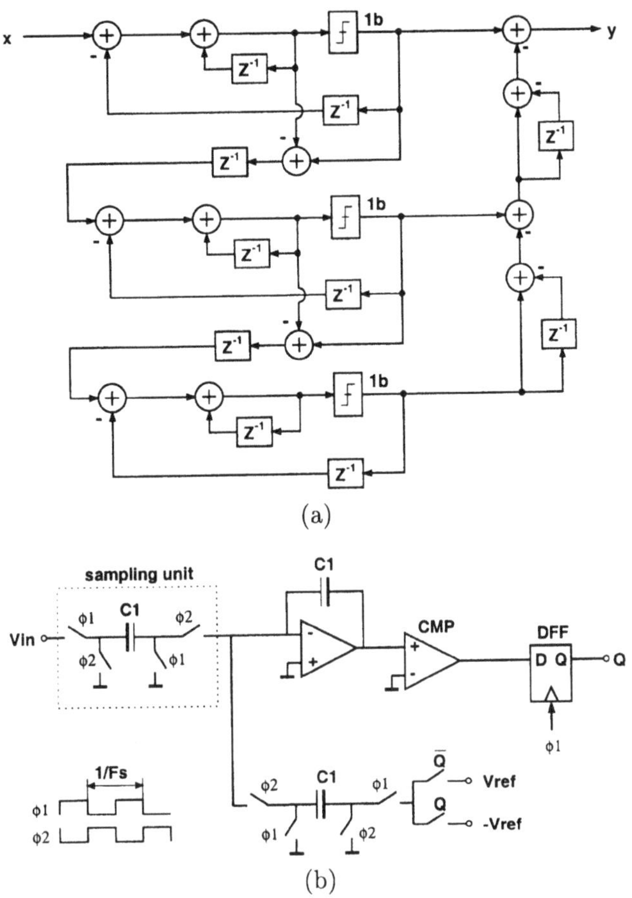

Figure 5.16: (a) System diagram for a third-order (1-1-1) cascaded ΔΣ modulator used in the simulations. (b) Circuit diagram of the first modulator. Single-ended version is shown for simplicity.

For the old DQS circuit, Figure 5.17(b) shows that the desired signal is down-converted to baseband at $-2^{-5}f_s$. However, due to the channel mismatch and phase error, a portion of the image interferer is folded to the baseband at $\pm 2^{-6}f_s$, as indicated by $I1$ and $I2$. Besides, a self-image, I_3, also appears in the baseband at $+2^{-5}f_s$. The I_1 and I_2 are suppressed by 46 dB in magnitude. The image rejection ratio is then equal to 40 dB which agrees with equation (5.22).

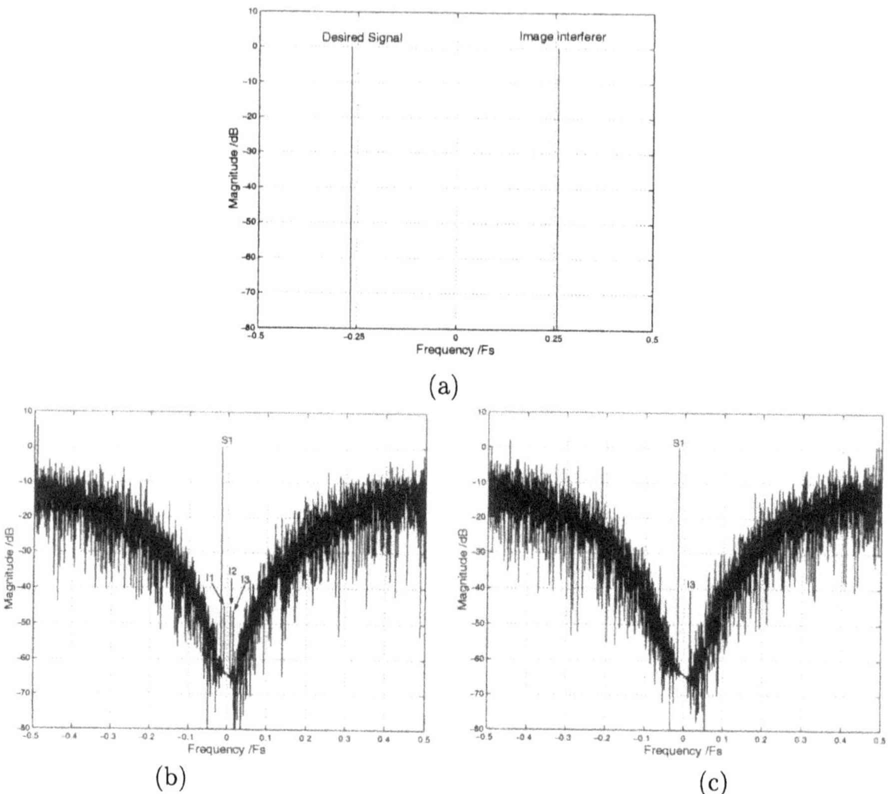

Figure 5.17: SWITCAP simulation results: (a) input spectrum; output spectrum of the (b) the old DQS scheme; (c) the proposed DQS scheme; Fixed capacitor mismatch of 1% and fixed clock phase error of 1° are assumed.

For the proposed DQS circuit, Figure 5.17(c) shows that the desired signal is also down-converted to baseband but there are no image components at $\pm 2^{-6} f_s$. Only the self-image I_3 at $+2^{-5} f_s$ exists. This proves our prediction of the mismatch and phase error insensitive property of the proposed DQS circuit. The self-image is suppressed about 40 dB in magnitude.

5.4.2 Double Image-Reject Sampling

To suppress the image interferers $n1$ and $n2$ in the analog DQS scheme illustrated in Figure 5.14, another method is to add an image rejection filter before the down-conversion of the signal. This concept is illustrated in the block diagram of Figure 5.18. Note that all blocks in Figure 5.18 operate in discrete-time domain. Since the DQS is an image suppressing scheme by itself, the proposed scheme is named as *double image-reject sampling.* It can be imagined that after being suppressed twice, the images $n1$ and $n2$ can be virtually completely removed even with moderate matching conditions.

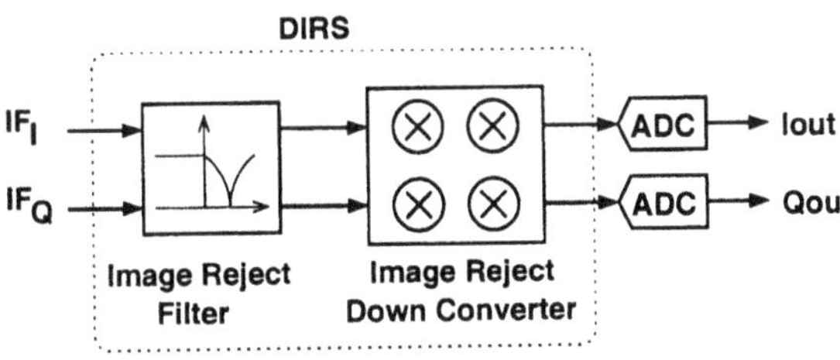

Figure 5.18: Architecture of double image-reject sampling scheme.

The image rejection filter, in this case, is a complex filter with notches around $\frac{f_s}{4}$. Remember that filters with this property are indeed Hilbert transformers. The design of discrete-time Hilbert transformers and switched-capacitor circuit realizations have been discussed in Chapter three. To process complex signals, the Hilbert transformer should be implemented in the form of complex filter as shown in Figure 5.19, where the $H_I(z)$ and $H_Q(z)$ consist of a Hilbert transformer pair.

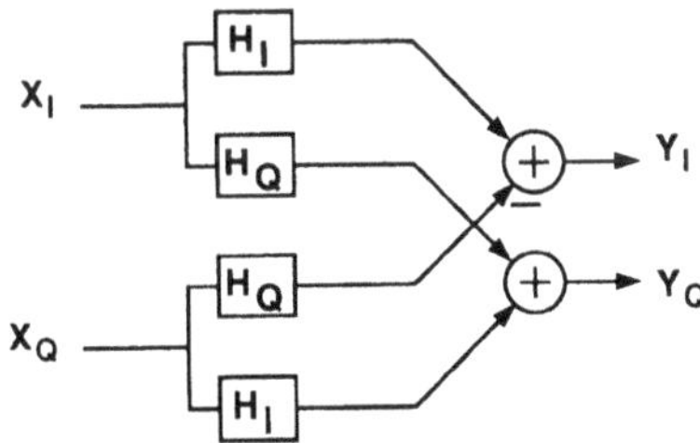

Figure 5.19: Architecture of a complex filter.

When oversampling A/D conversion approaches are used and if the oversampling ratio is large enough, then we can adopt a very simple image rejection filter - the first order FIR Hilbert transformer. This filter has a complex transfer

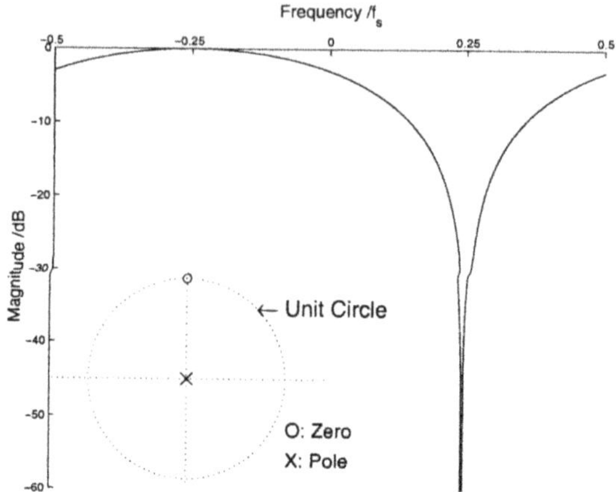

Figure 5.20: The magnitude response and zero-pole plot of the first order FIR Hilbert transformer.

function of $H(z) = H_I(z) + jH_Q(z)$, where

$$\begin{cases} H_I(z) = 1, \\ H_Q(z) = -z^{-1}. \end{cases} \tag{5.31}$$

The magnitude response $|H_I(z) + jH_Q(z)|$ of this complex filter has a notch at $f_s/4$ as shown in Figure 5.20. The corresponding zero-pole plot is also shown in Figure 5.20. The output Y_I and Y_Q of this filter can be written as:

$$Y_I = X_I + X_Q z^{-1} \tag{5.32}$$

$$Y_Q = X_Q - X_I z^{-1}. \tag{5.33}$$

The schematic diagram that combines this filter with the DQS scheme is shown in Figure 5.21.

SC Circuit

It is interesting that the DIRS with first order FIR image filter can be simply realized in the sampling units of I and Q A/D converters. The proposed circuit is shown in Figure 5.22, which is modified from Figure 5.12. In the original circuit of Figure 5.12, the switches s_1, s_2, s_3 and s_4 are connected to ground. In the new circuit, those switches are connected to corresponding input signals to realize the function of jz^{-1} required in the image rejection filter (5.31). This is

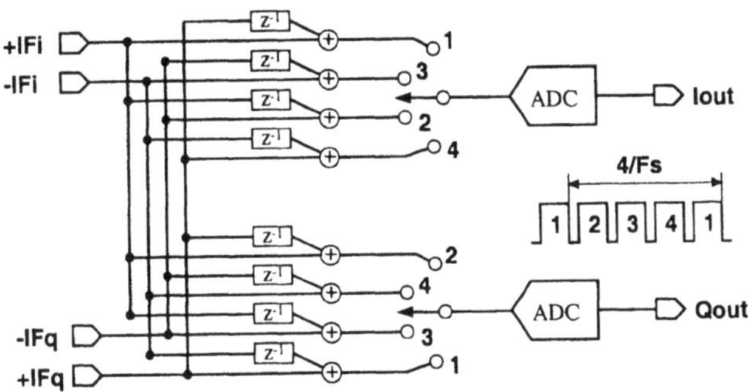

Figure 5.21: The schematic diagram of the DIRS with first-order FIR Hilbert transformer.

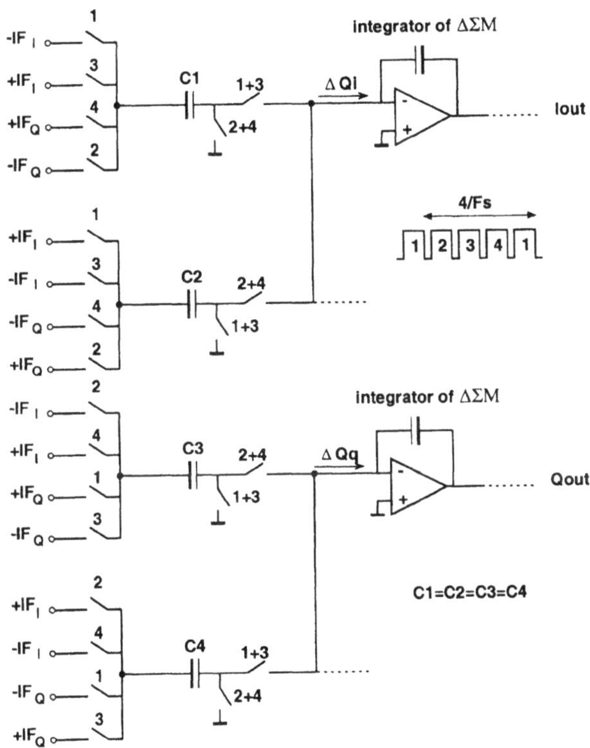

Figure 5.22: Switched-capacitor DIRS circuit.

the only difference between the two circuits. Just by this sampling arrangement, we can have achieve an additional rejection of those image interferers at $+\frac{f_s}{4}$. Comparing with the DQS circuit, the new circuit does not introduce any extra circuit elements or clock phases, i.e., no extra costs.

To understand the operation of the proposed circuit, let us first look at the charge ΔQ_i and ΔQ_q injected to the integrating capacitors of I and Q $\Delta\Sigma$ A/D converters respectively. At clock phase 1, 2, 3 and 4, they are

$$\begin{aligned}
\Delta Q_i[1] &= -C_1(V_{if,I} + V_{if,Q}z^{-1}), & \Delta Q_i[2] &= C_2(V_{if,Q} - V_{if,I}z^{-1}),\\
\Delta Q_i[3] &= C_1(V_{if,I} + V_{if,Q}z^{-1}), & \Delta Q_i[4] &= -C_2(V_{if,Q} - V_{if,I}z^{-1}),\\
\Delta Q_q[1] &= -C_3(V_{if,Q} - V_{if,I}z^{-1}), & \Delta Q_q[2] &= -C_4(V_{if,I} + V_{if,Q}z^{-1}),\\
\Delta Q_q[3] &= C_3(V_{if,Q} - V_{if,I}z^{-1}), & \Delta Q_q[4] &= C_4(V_{if,I} + V_{if,Q}z^{-1}),
\end{aligned}$$

where $V_{if,I}$ and $V_{if,Q}$ are I and Q input voltages respectively, C_1, C_2, C_3 and C_4 are the capacitances of sampling capacitors. Denote $P_I(z) = \mathcal{Z}\{[1\ 0\ -1\ 0]\}$ and $P_Q(z) = \mathcal{Z}\{[0\ 1\ 0\ -1]\}$, where $\mathcal{Z}$ is the $\mathcal{Z}$-transformation. The $\Delta Q_i(z)$ and $\Delta Q_q(z)$ can be written as:

$$\begin{aligned}
\Delta Q_i(z) &= -C_1(V_{if,I} + V_{if,Q}z^{-1}) \otimes P_I(z) + C_2(V_{if,Q} - V_{if,I}z^{-1}) \otimes P_Q(z),\\
\Delta Q_q(z) &= -C_3(V_{if,Q} - V_{if,I}z^{-1}) \otimes P_I(z) - C_4(V_{if,I} + V_{if,Q}z^{-1}) \otimes P_Q(z).
\end{aligned}$$

Assuming perfect capacitance matching, i.e., $C_1 = C_2 = C_3 = C_4 = C$, we obtain:

$$\begin{aligned}
&\Delta Q_i(z) + j\Delta Q_q(z)\\
&= -C(V_{if,I} + V_{if,Q}z^{-1} + j(V_{if,Q} - V_{if,I}z^{-1})) \otimes (P_I(z) + jP_Q(z))\\
&= -C(V_{if,I} + jV_{if,Q})(1 - jz^{-1}) \otimes (P_I(z) + jP_Q(z)).
\end{aligned} \tag{5.34}$$

Equation (5.34) says that the IF input $V_{if,I} + jV_{if,Q}$ is first filtered by the image-reject filter $1 - jz^{-1}$, and then mixed by $P_I(z) + jP_Q(z)$, before it injects to the integrator of $\Delta\Sigma$ modulator. Note that the image-reject filter used here is the complex conjugate of (5.31), because the desired frequency notch is at $\frac{1}{4}f_s$ instead of $-\frac{1}{4}f_s$ in this case.

Considering the presence of capacitor mismatches, i.e., $C_1 \neq C_2 \neq C_3 \neq C_4$ (the mismatch between the integrating capacitors of I and Q A/D Converters can be regarded as the mismatch between (C_1, C_2) and (C_3, C_4)), we obtain the negative charge injected to the integrating capacitors as:

$$\begin{aligned}
&-[\Delta Q_i(z) + j\Delta Q_q(z)]\\
= \;&(V_{if,I} + jV_{if,Q})(1 - jz^{-1}) \otimes P_A(z)\\
&+ \left[(V_{if,I} + jV_{if,Q})(1 - jz^{-1})\right]^* \otimes P_B(z)
\end{aligned} \tag{5.35}$$

where

$$P_A(z) = \frac{C_1 + C_4 + C_3 + C_2}{4} P(z) + \frac{C_1 - C_4 + C_3 - C_2}{4} P^*(z), \quad (5.36)$$

$$P_B(z) = \frac{C_1 + C_4 - (C_3 + C_2)}{4} P(z) + \frac{C_1 - C_4 - (C_3 - C_2)}{4} P^*(z). \quad (5.37)$$

and $P(z) = P_I(z) + jP_Q(z)$. The mismatch effect on DIRS scheme described in (5.35) is similar to the mismatch effect on the the DQS scheme described in (5.23). But in the former, the input $(V_{if,I} + jV_{if,Q})$ is first filtered by the image-reject filter $1 - jz^{-1}$, which is not affected by the capacitance mismatch as can be seen in (5.35).

Lastly, the non-even phase errors is not solved by the DIRS scheme because the erroneous sampling process happens before the image filtering process. However, this error can be well-controlled since the even and odd clocks have a frequency of only a half of f_s, i.e., the sampling frequency of the A/D converter. Alternatively, this problem can be avoided by employing in I and Q channels a sample-and-hold stage as shown in Figure 5.8, which is clocked at f_s.

Simulation Results

Firstly, high level simulations were conducted to verify the DIRS scheme by using Matlab^{TM} [35]. Frequency domain simulations based on equation (5.27) and (5.35) were carried out and the results are shown in Figure 5.23.

For perfect channel matching, Figure 5.23(a) shows the responses of both the old DQS and the DIRS systems to desired inputs which are located around $-f_s/4$. From the figure we find that the IF signals are translated to baseband at the output as expected. For 1% channel mismatch, Figure 5.23(b) shows the responses of DQS and DIRS to the desired input. The output in this case contains not only the baseband signal, but also the signal at $\pm f_s/2$ with magnitude of $-40dB$. The high frequency spurious response is generated by channel mismatches. These spurious signals do not matter because they are located at high frequency and will be filtered in the digital filter. From Figure 5.23(a) and (b), we can find that the DQS and DIRS have same responses to the desired signal.

Figure 5.23(c) and (d) show the responses of DQS and DIRS to the image interferer located around $f_s/4$, for 0% and 1% channel mismatches respectively. From Figure 5.23(c), we find no interferer folded to the baseband due to perfect channel matching. In the presence of channel mismatches, a portion of the interferer is folded to the baseband and corrupt the desired signal as shown in Figure 5.23(d). In DQS, the magnitude of the folded image interferer is $-40dB$ with respective to their original magnitude. But in DIRS, the interferer folded

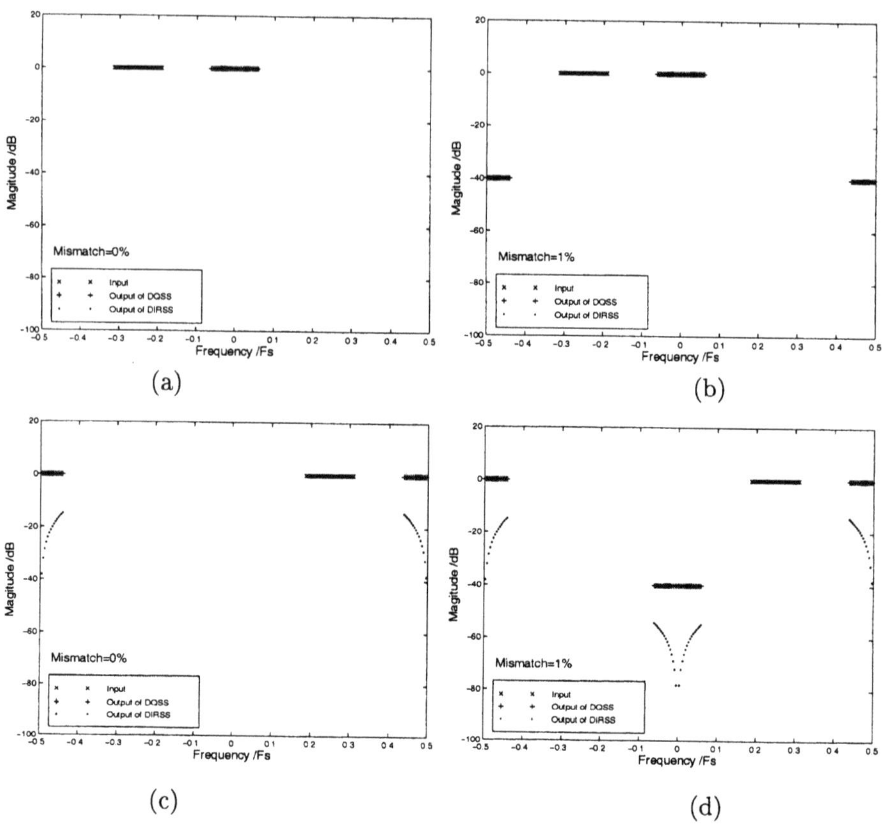

Figure 5.23: The responses of the old DQS circuit and the proposed DIRS circuit to the desired signal centered at $-f_s/4$ for (a) 0% and (b) 1% channel mismatch, and their responses to the image interferer centered at $f_s/4$ for (c) 0% and (d) 1% channel mismatch.

to the baseband is suppressed more than $-70dB$ because it is filtered by the complex notch filter $1 - jz^{-1}$ before down-conversion. This means that the image rejection ratio is increased by more than $30dB$.

Then, SWITCAP simulations have been conducted. The baseband $\Delta\Sigma$ modulator shown in Figure 5.16 is used again for A/D conversion. This times, the sampling unit shown in Figure 5.16(b) is replaced by the DIRS circuit shown in Figure 5.22.

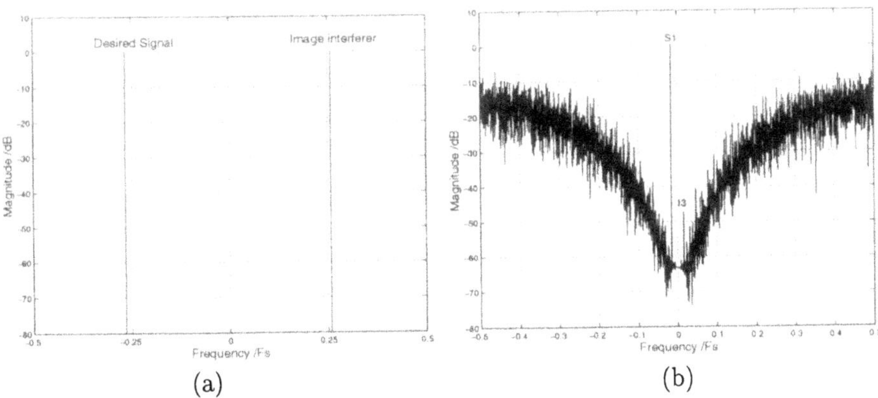

Figure 5.24: SWITCAP simulation results: (a) input spectrum, (b) output spectrum DIRS scheme. Fixed capacitor mismatch of 1% is assumed.

The same input tones shown in Figure 5.17(a) are applied as the inputs. With 1% capacitor mismatch and without phase error, the DIRS scheme gives an output as shown in Figure 5.24. In the output spectrum, no image components at $\pm 2^{-6} f_s$ is observed because they are lower than the noise level. The self-image is observed at $+2^{-5} f_s$ and it suppressed about 46 dB in magnitude. The result proves the super image rejection performance of the DIRS scheme under the presence of channel mismatches. However, it is worth to mention that the image rejection of DIRS scheme will be limited by clock phase errors.

5.5 Summary

In summary, image rejection improvement techniques for quadrature sampling in complex IF receivers have been presented in this chapter.

Quadrature sampling schemes in complex IF receivers can be divided into two classes. The first class refers to schemes with direct-IF digitising. The second class refers to schemes with IF-sampling and baseband-digitising. Characteristics of these quadrature sampling schemes are summarised in Table 5.2. If $\Delta\Sigma$ modulators are used for A/D conversion, then the first class needs bandpass shaping of quantisation noise, and the second class needs only lowpass shaping of quantisation noise. The existing sampling methods, listed in the second and fourth columns of Table 5.2, suffer from channel mismatch or non-even phase error and therefore have very limited image rejection performance.

Table 5.2: Characteristics of different quadrature sampling schemes for the complex IF receiver.

	Direct-IF digitising		IF-sampling BB-digitising		
	w/o image reject filter	w/ image reject filter	Old circuit Fig 5.12	New circuit Fig 5.15	DIRS Fig 5.22
ΣΔ ADC noise shaping	Bandpass	Bandpass	Lowpass	Lowpass	Lowpass
Channel mismatch effect#	$\frac{1}{(\Delta G/G)^2}$	$\frac{H(j\omega)}{(\Delta G/G)^2}$ †	$\frac{1}{(\Delta G/G)^2}$	No‡	$\frac{H(j\omega)}{(\Delta G/G)^2}$ †
Phase error effect#	No	No	$\frac{1}{(\theta/2)^2}$	No	$\frac{1}{(\theta/2)^2}$

† $H(j\omega) = 1 - je^{-jwt}$.
‡ Channel mismatch generate only self-image.
# in terms of IRR.

For IF-sampling and baseband-digitising, two approaches of improving the image rejection have proposed. The first approach is to add an first-order FIR complex notch filter in the IF stage. This filter is realized as a part of the sampling circuit therefore requires no extra cost. With this filter, the channel mismatch effect is greatly reduced, but the problem of non-even phase error is not solved. Second, a quadrature sampling circuit immune to both channel mismatch and phase error has been discovered. There is also no extra cost. This circuit should be widely used.

For direct-IF digitising, the approach for improving the image rejection is to add a first-order FIR complex notch filter in the IF stage. With two extra S/H units, the filter can be realized in the sampling unit of bandpass ΔΣ modulators. Again, this circuit should be widely used in direct-IF digitising complex-IF receivers.

References

[1] O.D. Grace and S.P. Pitt, "Quadrature sampling of high frequency waveforms," *J. Acoust. Soc. Amer.*, vol. 44, pp. 1453–1454, 1968.

[2] J. L. Brown, "On quadrature sampling of bandpass signals," *IEEE Trans.*

on Aerospace and Electronic Systems, vol. AES-15, no. 3, pp. 366–371, May 1974.

[3] W.M. Waters and B.R. Jarrett, "Bandpass signal sampling and coherent detection," *IEEE Trans. on Aerospace and Electronic Systems*, vol. AES-18, no. 4, pp. 731–736, Nov. 1982.

[4] V. Considine, "Digital complex sampling," *IEE Electronics Letters*, vol. 19, no. 16, pp. 608–609, 4th Aug. 1983.

[5] Charles M. Rader, "A simple method for sampling in-phase and quadrature components," *IEEE Trans. on Aerospace and Electronic Systems*, vol. AES-20, no. 6, pp. 821–824, Nov. 1984.

[6] R.L. Mitchell, "Creating complex signal samples from a band-limited signal," *IEEE Trans. on Aerospace and Electronic Systems*, vol. 25, no. 3, pp. 425–427, May 1989.

[7] H. Liu, A. Ghafoor, and P.H. Stockmann, "A new quadrature sampling and processing approach," *IEEE Trans. on Aerospace and Electronic Systems*, vol. 25, no. 5, pp. 733–747, Sept. 1989.

[8] L.E. Pellon, "A double Nyquist digital product detector for quadrature sampling," *IEEE Trans. on Signal Processing*, vol. 40, no. 7, pp. 1670–1681, July 1992.

[9] V. Eerola, T. Ritoniemi, and H. Tenhunen, "Second-order sampling and oversampled A/D and D/A converters in digital data transmission," in *Proc. IEEE Int. Symposium on Circuits and Systems*, 1991, vol. 3, pp. 1521–1524.

[10] H. Tao and J.M. Khoury, "Direct-conversion bandpass $\Delta\Sigma$ modulator," *Electronics letters*, vol. 33, no. 5, pp. 1282–1283, July 17th 1997.

[11] H. Tao and J.M. Khoury, "A 100MHz IF, 400 MSample/s CMOS direct conversion bandpass $\Delta\Sigma$ modulator," in *Digest of Technical Papers, IEEE Int. Solid-State Circuit Conference*, Feb. 1999, pp. 60–61.

[12] Darwin T.S. Yim and Curtis C. Ling, "A 200-MHz CMOS I/Q down-converter," *IEEE Trans. Circuit and Systems - II*, vol. 46, no. 6, June 1999.

[13] S. Lindfors, M. L ansirinne, and K. Halonene, "A 2.7V 50MHz IF-sampling DS-modulator with +37dBV IIP_3 for digital cellular phones," in *Proc. IEEE European Solid-State Circuits Conference*, 2000.

[14] T. Burger and Q. Huang, "A 13.5mW, 185MSample/s $\Delta\Sigma$-modulator for UMTS/GSM dual-standard IF reception," in *Digest of Technical Papers, IEEE Int. Solid-State Circuit Conference*, Feb. 2001, pp. 44–45.

[15] C. Azeredo Leme, Ricardo Reis, and Eduardo Viegas, "Wideband subsampling A/D conversion with image rejection," in *IEEE Workshop on Wireless Communication Circuits and Systems*, Lucerne, Switzerland, June 1998.

[16] D.W. Rice and K.H. Wu, "Quadrature sampling with high dynamic range," *IEEE Trans. on Aerospace and Electronic Systems*, vol. AES-18, no. 4, pp. 736–739, Nov. 1982.

[17] S. Jantzi, R. Schreier, and M. Snelgrove, "The design of bandpass $\Delta\Sigma$ ADCs," in *Delta-Sigma Data Converters, Theory, Design and Simulation*, S. Norsworthy, R. Schreier, and G.C. Temes, Eds., pp. 282–308. IEEE Press, 1997.

[18] H.J. Dressler, "Interpolative bandpass A/D conversion - experimental results," *IEE Electron. Letters*, vol. 26, no. 20, pp. 1652–1653, Sept. 1990.

[19] A.M. Thurston, T.H. Pearce, and M.J. Hawksford, "Bandpass implementation of the sigma-delta A-D conversion technique," *Proc. IEE Int. Conference on A/D and D/A Conversion, Swansea, U.K.*, pp. 81–86, Sept. 1991.

[20] S.A. Jantzi, W.M. Snelgrove, and P.F. Ferguson Jr., "A fourth-order bandpass sigma-delta modulator," *IEEE J. Solid-State Circuits*, vol. 28, no. 3, pp. 282–291, March 1993.

[21] G. Tröster et al., "An interpolative bandpass converter on a 1.2μm BiCMOS analog/digital array," *IEEE J. Solid-State Circuits*, vol. 28, no. 4, pp. 471–477, April 1993.

[22] L. Longo and B.R. Horng, "A 15b 30kHz bandpass sigma-delta modulator," in *Digest of Technical Papers, IEEE Int. Solid-State Circuit Conference*, 1993, pp. 226–227.

[23] J.E. Eklund and R. Arvidsson, "A multiple sampling, single A/D conversion technique for I/Q demodulation in CMOS," *IEEE J. Solid-State Circuits*, vol. 31, no. 12, pp. 1987–1994, Dec. 1996.

[24] J. A. C. Bingham, "Applications of a direct-transfer SC integrator," *IEEE Trans. Circuits and Systems*, vol. CAS-31, pp. 419–420, 1984.

[25] J.C. Rudell, J.J. Ou, et al., "A 1.9GHz wide-band IF double conversion CMOS receiver for cordless telephone application," *IEEE J. Solid-State Circuits*, vol. 32, pp. 2071–2088, Dec 1997.

[26] D.K. Weaver, "A third method of generation and detection of single-sideband signals," *Proc. IRE*, vol. 44, pp. 1703–1705, Dec 1956.

[27] P. Aziz, H. Sorensen, and J. Van der Spiegel, "Performance of complex noise transfer functions in bandpass and multi-band sigma delta systems," in *Proc. IEEE Int. Symposium on Circuits and Systems*, 1995, pp. 641–644.

[28] S.A. Jantzi et al., "Complex bandpass $\Sigma\Delta$ converter for digital radio," in *Proc. IEEE Int. Symposium on Circuits and Systems*, May-June 1994, vol. 5, pp. 453–456.

[29] S.A. Jantzi, K.W. Martin, and A. S. Sedra, "Quadrature bandpass $\Delta\Sigma$ modulation for digital radio," *IEEE J. Solid-State Circuits*, vol. 32, no. 12, pp. 1935–1950, Dec. 1997.

[30] A. Swaminathan, "A single-IF receiver architecture using a complex sigma-delta modulator," M.S. thesis, Carleton University, Ottawa, Canada, 1997.

[31] K.P. Pun, J.E. Franca, and C. Azeredo Leme, "A quadrature sampling scheme with improved image rejection for complex IF receivers," in *IEEE Int. Symposium on Circuits and Systems*, Sydney, Australia, May 2001.

[32] D. G. Haigh and B. Singh, "A switching scheme for switched capacitor filters which reduces the effect of parasitic capacitances associated with switch control terminals," in *Proc. IEEE Int. Symposium on Circuits and Systems*, 1983, vol. 2, pp. 586–589.

[33] K. Suyama and S.C. Fang, *User's Manual for SWITCAP2 version 1.1*, Columbia University, 1992.

[34] Y. Matsuya, K. Uchimura, A. Iwata, et al., "A 16-bit oversampling A-to-D conversion technology using triple-integration noise shaping," *IEEE J. Solid-State Circuits*, vol. 22, pp. 921–929, Dec. 1987.

[35] The Math Works Inc., *User's guide, MATLAB 5*, 1996.

[36] K.P. Pun, J.E. Franca, and C. Azeredo Leme, "A switched-capacitor image rejection filter for complex IF receivers," in *Proc. IEEE Asia-Pacific Conference on Circuits and Systems*, Tianjin, China, Dec. 2000.

Chapter 6

Digital Calibration of I/Q Mismatches

6.1 Introduction

In wireless receivers that employ analog quadrature demodulation, analog quadrature sampling or double quadrature sampling schemes as described in the previous chapter, gain and phase imbalances between I and Q channels always exist. The I/Q mismatch will produce the so-called "self-image" problem, that is, the folding of the positive frequency components of the received signal to the negative frequency range and vice versa. In a receiver which receives only one radio channel signal, the impact of the self-image is the lowering of the signal-to-noise ratio [1]. If I/Q mismatches are not severe, say, amplitude mismatch below 1 dB and phase error below $5°$, the self-image is usually tolerable in such a single channel receiver.

On the other hand, if a receiver is designed to accommodate multi-channel signals, then the problem of I/Q mismatches can be very problematic as explained in the next paragraph. This situation occurs when the channel selection function of the receiver is brought from the analog to the digital domain to allow multi-mode operations. An example is the FM/AM dual mode receiver described in [2]. In the receiver, the same A/D converters are used for both FM and AM receptions. When it operates in AM mode, the receiver must accommodate multiple AM channels due to the relatively narrow AM bandwidth. Such situation can also occur in other multi-mode receivers, for example, a GSM/UMTS receiver. Other examples of multi-channel receiver can be found in base stations [3] of wireless communication systems.

The effect of I/Q mismatches in a multi-channel receiver is no longer to generate a self-image, but to generate image components of all the signals within the receiving band, as illustrated in Fig. 6.1. Since within the receiving band weak signals and strong signals can co-exist and a strong signal can be the image of a weak signal, even a small amount of I/Q mismatches can lead to the loss of the weak signal. Therefore, the I/Q mismatches are very problematic in a multi-channel receiver.

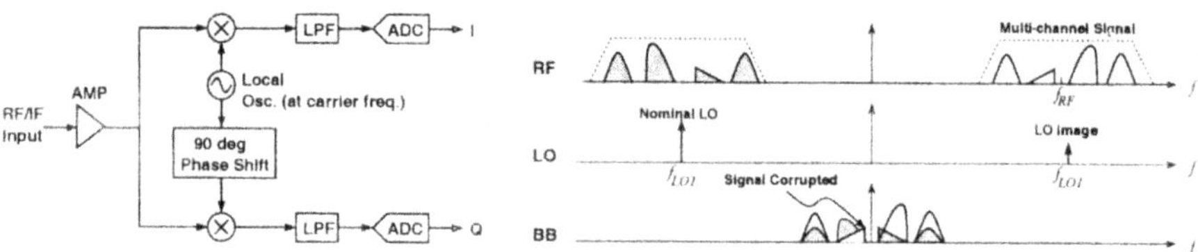

Figure 6.1: (a) Architecture of an analog quadrature receiver; (b) I/Q mismatch effect.

Methods exist in the literature to calibrate the I/Q mismatches by both analog and digital techniques. In this chapter, we will focus on the digital techniques.

Twenty years ago, Churchill *et al* developed a standard deterministic technique which injects a local sinusoidal test signal [4] and uses Gram-Schmidt procedure to correct the I/Q errors. Ten years later, MacLeod [5] proposed a similar method with a detailed description of the calibration procedure. Recently, Green *et al* introduced a nonlinear regression technique [6] to estimate the I/Q errors, which also applies a sinusoidal test signal. Kulkarni and Kostinski gave a simple method of I/Q error monitoring by using the received data itself or by injection of an arbitrary test signal [7]. Similar calibration techniques can be found in pipelined or parallel analog-to-digital converters [8, 9, 10]. All the methods mentioned above assume frequency-independent I/Q mismatches. If the bandwidth of the receiver is narrow, then the assumption is valid within the band of interest. For I/Q calibration of wideband receivers, Lee demonstrated a calibration scheme in [11]. However, it is only useful for the radar receivers that process only a single frequency signal at a time.

In this chapter, a wideband calibration method which can correct the frequency dependent mismatches is developed. It involves a mismatch measurement process by local test signal injection and an error correction process, all implemented in baseband digital domain. For the error measurement process, both the digital Fourier transform (DFT) and statistical estimation methods can be used. The statistical method used here is made compatible with the DFT method.

This chapter is arranged as follows. First, we compares the analog and digital calibrations. Then, two narrow-band I/Q mismatch calibration methods, namely, the Churchill's and the statistical methods are presented. And then the calibration method is extended to wideband receivers.

6.2 Analog Calibration vs Digital Calibration

There are both analogue (pre-ADC) and digital (post-ADC) options for calibrating the I/Q mismatches. Figure 6.2 shows the conceptual diagram of both methods. For the analog approach, we first measure the I/Q errors at output, then generate a proper analogue quantity, and subtract it from the analogue input. Adjustable gain stages and phase shifters are necessary circuit components in this approach.

Digital calibration methods that employ no feedback path to the analogue circuits involve only the processing of digital output as shown in Figure 6.2(b). Comparing with analogue methods, digital methods have the following advantage: 1) High accuracy is straightforward; 2) Added noise is not a concern; 3) Adjustment is immediate (no settling time), and 4) Easy for VLSI implementation.

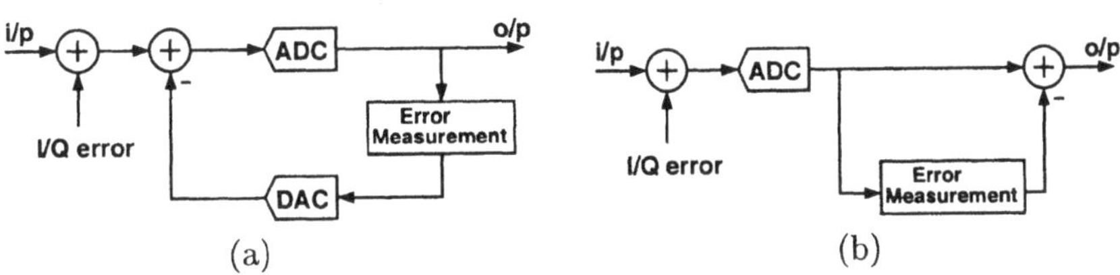

Figure 6.2: Conceptual diagram of (a) analog and (b) digital I/Q error correction.

6.3 Churchill's Method

In this section, a digital calibration of I/Q mismatches for narrow-band system is presented. By narrow-band system, we mean that the I/Q mismatches are constant over the band of interest.

Consider a quadrature receiver with a sinusoidal input. With the input, the quadrature output pair of the receiver can be expressed as:

$$I_1 = (1+\alpha)A\cos(2\pi f_i t) + i_{dc} \quad (6.1)$$

$$Q_1 = A\sin(2\pi f_i t + \epsilon) + q_{dc} \quad (6.2)$$

where A and f_i are signal amplitude and frequency respectively, α and ϵ represents the gain and phase imbalances respectively, and i_{dc} and q_{dc} are DC errors for I and Q channels. The error assignment is arbitrary, but since the errors are relative errors, there is no loss of generality.

F.E. Churchill proposed using the Gram-Schmidt orthogonalisation procedure to correct the I/Q errors as follows [4]:

$$\begin{bmatrix} I_2 \\ Q_2 \end{bmatrix} = \begin{bmatrix} 1+E & 0 \\ P & 1 \end{bmatrix} \begin{bmatrix} I_1 - i_{dc} \\ Q_1 - q_{dc} \end{bmatrix} \quad (6.3)$$

where

$$E = \frac{\cos\epsilon}{1+\alpha} - 1, \quad (6.4)$$

$$P = -\frac{\sin\epsilon}{1+\alpha}. \quad (6.5)$$

From (6.1) to (6.5), we have

$$I_2 = A\cos(2\pi f_i t)\cos\epsilon, \quad (6.6)$$

$$Q_2 = A\sin(2\pi f_i t)\cos\epsilon. \quad (6.7)$$

Therefore, the processed signals I_2 and Q_2 are orthogonal to each other and of equal amplitude, which means I/Q mismatches have been removed. Figure 6.3 shows such a mismatch correction circuit.

If the I/Q mismatches (α and ϵ) are known, then the parameters E and P are also known and the mismatches can be corrected by the operation presented above. But in general, α and *epsilon* are unknown. A DFT-based method to find E and P is employed in Churchill's method as presented below.

First, a sinusoidal test signal at a quarter of the sampling frequency of the ADC above the local oscillator frequency is injected to the I/Q demodulator. It will be clear later that the test signal at this frequency is the best choice because it requires minimum DFT length while can still provide DC offset, gain and phase mismatch information.

Second, we wait for the ADC outputs to settle, then collect a number of four-sample cycles of the ADC output to take the average. Denote $\mathbf{I}_1$ and $\mathbf{Q}_1$

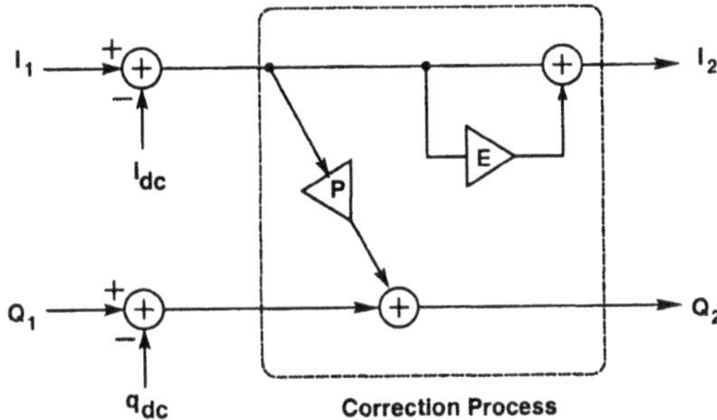

Figure 6.3: The narrow-band I/Q mismatch correction circuit.

as the averaged four-sample output vectors of I and Q respectively. A 4-point DFT is performed on these data:

$$\tilde{s}_1 = DFT\{\mathbf{I}_1 + j\mathbf{Q}_1\} = \begin{bmatrix} 1 & 1 & 1 & 1 \\ 1 & -j & -1 & j \\ 1 & -1 & 1 & -1 \\ 1 & j & -1 & -j \end{bmatrix} \mathbf{s}_1, \tag{6.8}$$

where $\mathbf{s}_1 = \mathbf{I}_1 + j\mathbf{Q}_1$.

The first component of $\tilde{s}_1$,i.e., $\tilde{s}_1[0]$ gives the estimated DC offset values:

$$\hat{i}_{dc} = \frac{1}{4}Re\{\tilde{s}_1[0]\}, \tag{6.9}$$

$$\hat{q}_{dc} = \frac{1}{4}Im\{\tilde{s}_1[0]\}. \tag{6.10}$$

Note that large DC offsets must be removed before the ADC to avoid saturating the analog circuits.

The I/Q error correction parameters E and P are estimated from the DFT output (see Appendix 6.A for proof):

$$\hat{E} = -Re\{\frac{2\tilde{s}_1[3]}{\tilde{s}_1^*[1]+\tilde{s}_1[3]}\} \tag{6.11}$$

$$\hat{P} = -Im\{\frac{2\tilde{s}_1[3]}{\tilde{s}_1^*[1]+\tilde{s}_1[3]}\}, \tag{6.12}$$

where * denotes complex conjugate.

6.3.1 Calibration with test frequency other than $f_s/4$

In some situations, the test signal frequency may differ from $f_s/4$ as described above. However, the calibration method still applies with a little modification.

Let f_i be the test signal frequency minus the LO frequency, f_s be the sampling frequency of ADCs. Let $f_i/f_s = F/M$, where F and M are relative prime numbers. The length of DFT is no longer 4. It is chosen as an integer multiple of $M \times F$ (or simply $M \times F$) to avoid truncation error in the DFT. For example, if $f_i = \frac{3}{8} f_s$, we should choose DFT length to be 24.

Now, the signal lies in the $(F^2 + 1)$th pin of the DFT (DC is the first pin), and the image lies in the $(M \times F - F^2 + 1)$th pin. Therefore, in calculating P and E, $\tilde{s}_1[1]$ and $\tilde{s}_1[3]$ in (6.11) (6.12) should be generalised to $\tilde{s}_1[F^2]$ and $\tilde{s}_1[M \times F - F^2]$ respectively.

6.3.2 Residual Image

The I/Q errors can be completely corrected if the system is free of noise. However thermal noise and flicker noise from the mixers, phase noise from the LO and the quantisation noise from the ADC all contribute to the output. In addition, the test signal itself is accompanied by noise. These noises will produce a residual error after the calibration. In the following paragraphs we will show that the residual image power increases with the SNR at the ADC output and decreases with the degree of filtering or averaging used in deriving the coefficients E and P.

The corrected signals I_2 and Q_2 of Figure 6.3 can be written as

$$\begin{aligned} I_2(t) &= (1+\hat{E})(1+\alpha)A\cos(2\pi f_i t) + (1+\hat{E})(i_{dc} - \hat{i}_{dc}) \quad (6.13) \\ Q_2(t) &= A\sin(2\pi f_i t + \epsilon) + \hat{P}(1+\alpha)A\cos(2\pi f_i t) \\ &\quad + \hat{P}(i_{dc} - \hat{i}_{dc}) + q_{dc} - \hat{q}_{dc}. \quad (6.14) \end{aligned}$$

Performing the 4-point DFT on $(I_2 + jQ_2)$, we get the residual image as:

$$v_{img} \equiv \tilde{s}_2[3]/4 = (A/2)\{(1+\hat{E})(1+\alpha) - \cos\epsilon + j[\sin\epsilon + \hat{P}(1+\alpha)]\}. \quad (6.15)$$

And the the image power is:

$$\begin{aligned} v_{img}^2 &= (A^2/4)[(1+\alpha)^2 + 2\hat{E}(1+\alpha)^2 + \hat{E}^2(1+\alpha)^2 + 1 - 2(1+\alpha)\cos\epsilon \\ &\quad - 2\hat{E}(1+\alpha)\cos\epsilon + \hat{P}^2(1+\alpha)^2 + 2\hat{P}(1+\alpha)\sin\epsilon]. \quad (6.16) \end{aligned}$$

The average power or the mean-squared value of the image is

$$\begin{aligned} \overline{v_{img}^2} &= (A^2/4)[(1+\alpha)^2 + 2\overline{\hat{E}}(1+\alpha)^2 + \overline{\hat{E}^2}(1+\alpha)^2 + 1 - 2(1+\alpha)\cos\epsilon \\ &\quad - 2\overline{\hat{E}}(1+\alpha)\cos\epsilon + \overline{\hat{P}^2}(1+\alpha)^2 + 2\hat{P}(1+\alpha)\sin\epsilon], \quad (6.17) \end{aligned}$$

where $\overline{\bullet}$ represents the ensemble mean.

By definition the mean-squared value of a random variable is equal to the sum of the variance and the square of the mean, $\overline{\hat{E}^2} = \sigma^2_{\hat{E}} + (\overline{\hat{E}})^2$, $\overline{\hat{P}^2} = \sigma^2_{\hat{P}} + (\overline{\hat{P}})^2$. From (6.4) and (6.5), $\overline{\hat{E}} = \cos\epsilon/(1+\alpha) - 1$, $\overline{\hat{P}} = -\sin\epsilon/(1+\alpha)$. Substituting these values to (6.17), we have

$$\overline{v^2_{img}} = [A^2(1+\alpha)^2/4](\sigma^2_{\hat{E}} + \sigma^2_{\hat{P}}). \tag{6.18}$$

Now let us find the variance of each of the correction coefficients $\hat{E}$ and $\hat{P}$ in terms of the input noise σ^2. To include the noise, we can express the test signal as:

$$s(t) = (1+\alpha)A\cos(2\pi f_i t) + i_{dc} + n_x(t) + j[A\sin(2\pi f_i t + \epsilon) + q_{dc} + n_y(t)], \tag{6.19}$$

where $n_x(t)$ and $n_y(t)$ are each Gaussian noise with variance $(1+\alpha)^2\sigma^2$ and σ^2, respectively.

From (6.11) and (6.12), the sum of $\sigma^2_{\hat{E}}$ and $\sigma^2_{\hat{P}}$ is equal to the noise in $\tilde{s}_1[3]$ scaled by $4/(\tilde{s}_1^*[1] + \tilde{s}_1[3])^2$. The noise in $\tilde{s}_1[3]$ is simply the sum of N in-phase and N quadrature samples normalised by $1/N^2$:

$$\sigma^2_{\tilde{s}_1[3]} = (1/N^2)[N(1+\alpha)^2\sigma^2 + N\sigma^2] = (\sigma^2/N)[(1+\alpha)^2 + 1], \tag{6.20}$$

where N is the total number of samples used in the DFT. The scale factor from (6.11) and (6.12) is

$$4/(\tilde{s}_1^*[1] + \tilde{s}_1[3])^2 = 4/A^2(1+\alpha)^2. \tag{6.21}$$

From the above equations, the sum of the variance of $\hat{E}$ and $\hat{P}$ is found to be

$$\sigma^2_{\hat{E}} + \sigma^2_{\hat{P}} = 2\sigma^2[(1+\alpha)^2 + 1]/[N(1+\alpha)^2A^2]. \tag{6.22}$$

Substituting (6.22) to (6.18), the mean-squared value of the residual image after the calibration is found to be

$$\overline{v^2_{img}} = \sigma^2[(1+\alpha)^2 + 1]/N. \tag{6.23}$$

The mean value of v_{img} is zero and hence the variance of the residue image, σ^2_{img}, is

$$\sigma^2_{img} = [1 + (1+\alpha)^2](\sigma^2/N) \cong 2\sigma^2/N \text{ for } \alpha \ll 1. \tag{6.24}$$

Therefore the image rejection ratio after the calibration is N/SNR, where SNR is the signal-to-noise ratio at the output of I and Q ADCs and N is the number of samples used in the DFT. Note that the relation that the complex output signal power is the sum of the powers of I and Q outputs has been used. From the above equation, perfect IRR can be achieved if N is sufficient large.

6.4 Statistical Method

Another way to estimate the correction parameters E and P based on the statistical independence of the I/Q signals is introduced here.

Refer to Figure 6.3, DC offsets i_{dc}, q_{dc} must be subtracted from the outputs firstly. They are the time average of I and Q outputs: $i_{dc} =< I >$, $q_{dc} =< Q >$, where $< \bullet >$ is the time averaging operator. A circuit to estimate the DC offset is shown in Figure 6.4. In Figure 6.4 the subtracter, the gain and the accumulator form a negative feedback loop. If there is a DC component, the accumulator will accumulate this error until the negative feedback forces it approaching zero, provided that the step size μ is small enough to eliminate the variation in the output of the accumulator.

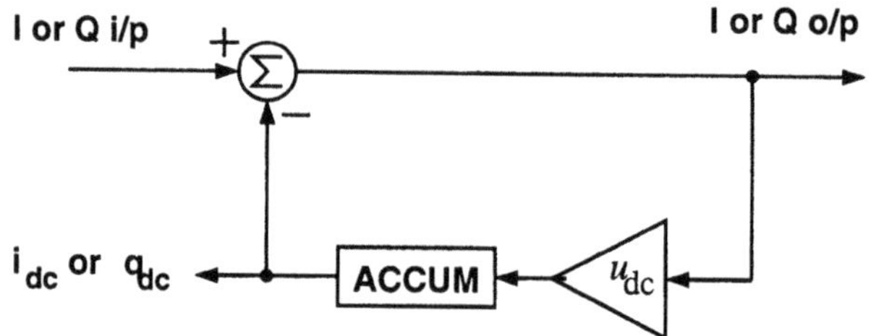

Figure 6.4: DC error correction, where μ_{dc} is a step size.

For the phase error, the circuit shown in Figure 6.5(a) can be used to estimate the correction parameter P [12]. The circuit is indeed a decorrelation system. In Figure 6.5(a) the output of the accumulator is the accumulation of the product of I and Q, which represents their cross-correlation. If there is a phase error, the cross-correlation will be non-zero, and the accumulator will accumulate this error until the negative feedback loop forces it approaching zero (therefore decorrelate I and Q signals).

This phase correction system uses the statistical property that the I and Q signals are uncorrelated in the absence of phase error. The phaser diagram of Figure 6.5(b) illustrate the phase correction process.

In principle, this system does not need a special calibration signal. The received signal itself can be used as the calibration signal [13] as long as their are uncorrelated in the absence of the phase error. Any deterministic signal which can be expressed as a Fourier series has uncorrelated I and Q components. However, if we want to calibrate the error at a specific frequency, then we need a sinusoidal test signal.

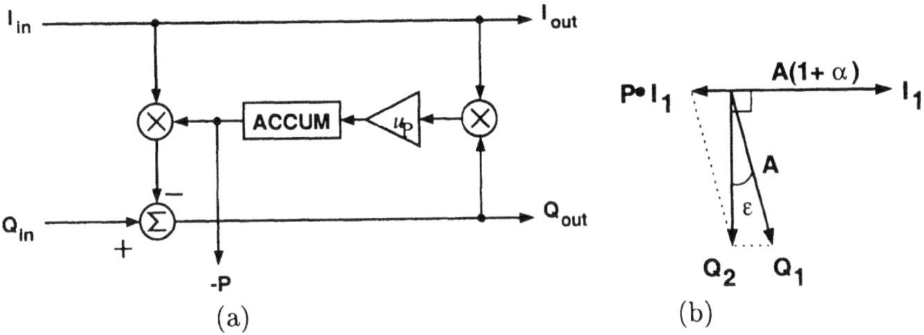

Figure 6.5: (a) Phase error correction system. (b)Phaser diagram. After correction $\vec{Q}_2 = \vec{Q}_1 + P\vec{I}_1$ is orthogonal to $\vec{I}_1$.

For I and Q input given by (6.1) (6.2), the output P of Figure 6.5 is given by (from (6.59) in Appendix 6.B):

$$P = -\frac{\sin\epsilon}{1+\alpha}\frac{1}{1+1/\mathrm{SNR}} \tag{6.25}$$

where SNR is the signal-to-noise ratio at the output of the ADCs of the receiver and is equal to $A^2/2\sigma^2$ and σ^2 is the variance of the noise at ADC output. We see that P is very close to $-\frac{\sin\epsilon}{1+\alpha}$ except for very low SNR. Also, P equals zero when no phase error presents.

For the gain error, the system shown in Figure 6.6 can be used to estimate the parameter E. In this system, the accumulator accumulates the amplitude difference of I and Q, and then the amplitude of I is adjusted until it is equal to that of Q.

For I and Q input given by (6.1) (6.2), we have the output E of Figure 6.6 as (from (6.66) in Appendix 6.B):

$$\begin{aligned} E &= \frac{\cos\epsilon}{1+\alpha}\sqrt{1+\frac{1+2\mathrm{SNR}}{(1+\mathrm{SNR})^2}\tan^2\epsilon} - 1 \\ &\approx \frac{\cos\epsilon}{1+\alpha} - 1, \quad \text{for } \epsilon \ll 1. \end{aligned} \tag{6.26}$$

From the above equation, the estimation of E is less affected by noise.

The correction systems presented above must be placed in a proper order: first DC error correction, then phase error and finally the gain error. It is obvious that the DC error must be corrected before phase and gain error correction. In the phase correction process, an additional gain error is introduced. While

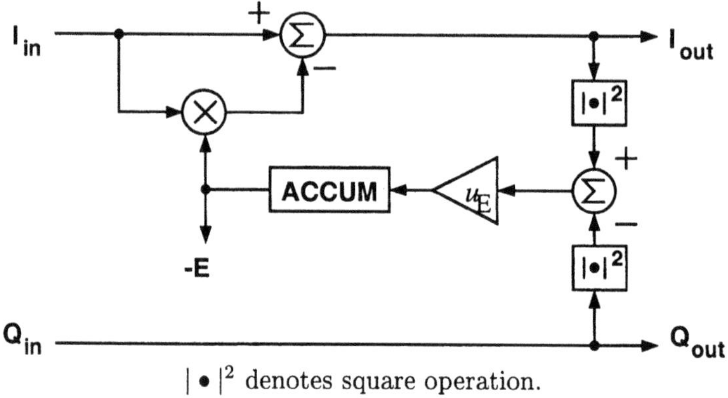

$| \bullet |^2$ denotes square operation.

Figure 6.6: Gain mismatch correction.

in the gain correction process, the phases of I and Q are not affected. Therefore the phase correction must be placed before the gain correction.

6.4.1 Residual image

It can be easily shown that the image power is related to the phase error ϵ and gain error α as

$$v_{img}^2 = \frac{A^2}{4}(\epsilon^2 + \alpha^2), \tag{6.27}$$

where A is the signal amplitude. From (6.63) and (6.69) in Appendix 6.B, the residual phase error ϵ' and gain error α' after the processing are given by

$$\epsilon' = -\frac{\tan \epsilon}{1 + \text{SNR}}, \tag{6.28}$$

$$\alpha' = \frac{\tan^2 \epsilon}{\text{SNR}}. \tag{6.29}$$

The α' is much less than ϵ', so the residual image is mainly caused by ϵ'. From (6.27) and (6.28), we have the residual image power as

$$\frac{A^2}{4}\frac{\tan^2 \epsilon}{(1 + \text{SNR})^2} \cong \frac{A^2}{4}\frac{\epsilon^2}{\text{SNR}^2} \quad \text{for } \epsilon \ll 1 \text{ and SNR} \gg 1. \tag{6.30}$$

Suppose the phase error is the dominant factor, from (6.27) and (6.30), we find that the image power after the calibration is 2SNR (in decibels) lower than before the calibration .

6.4.2 Convergence speed and stability

The phase and gain correction shown in Figure 6.6 and Figure 6.5 involves nonlinear operations: multiplication and squaring. A simply analytical solution does not exist. However through a proper representation of the error generation and correction they can be analysed by a standard linear system method. The detailed analysis is given in Appendix 6.C.

From Appendix 6.C, the convergence time constant of P is:

$$\tau_P = \frac{2}{\mu_P A^2}, \tag{6.31}$$

where τ_P is the number of samples needed for P to reach $(1 - e^{-1})$ of its steady-state value, A is the amplitude of the test signal and μ_P is the step size.

The convergence time constants for gain and DC error corrections are given by

$$\tau_E = \frac{1}{\mu_E A^2} \quad \text{and} \tag{6.32}$$

$$\tau_{dc} = \frac{1}{\mu_{dc}} \tag{6.33}$$

respectively.

In the analysis earlier this section, we assumed P a constant. The value of P given by (6.25) is actually a statistical mean value. Apart from that, P contains also a noise component due to the integration of noises in the feedback loop. At the end of calibration process, the variance of P is:

$$\begin{aligned} \sigma_P^2 &= \mu_P T_{cal} \sigma_{n_I}^2 \sigma_{n_Q}^2 \\ &= \mu_P T_{cal} (1+\alpha) \sigma^4 \\ &\cong \mu_P T_{cal} \sigma^4 \end{aligned} \tag{6.34}$$

where T_{cal} is the period of calibration in terms of samples. If we choose $T_{cal} = 7\tau_P$ such that P equals 99.9% of its final value, then the variance of P becomes

$$\sigma_P^2 = \mu_P \frac{14}{\mu_P A^2} \sigma^4 = \frac{14}{A^2} \sigma^4. \tag{6.35}$$

As σ^4 is very small, the deviation of P from its nominal value can be neglected. Besides, it is interesting to see that the variance of P does not depend on the step size.

Lastly, the DC error, gain and phase correction are absolutely stable because they are one-pole systems.

6.4.3 Simulation Results

Examples and simulation results about the DFT and statistical calibration methods are discussed in this subsection.

The simulation is based on the quadrature receiver model with I/Q imbalances as shown in Figure 6.7. In the model, $s(t)$ and LO are the input and local oscillator. α, ϵ, and i_{dc}, q_{dc} are gain, phase and DC errors respectively. n_I and n_Q are zero-mean Gaussian noise ($\sigma_{n_I} = (1+\alpha)\sigma_{n_Q}$). As mentioned before, the SNR is defined as $A^2/2\sigma_{n_Q}^2$, where A is the signal amplitude at ADC output. The simulation is carried out on Matlab Simulink[14].

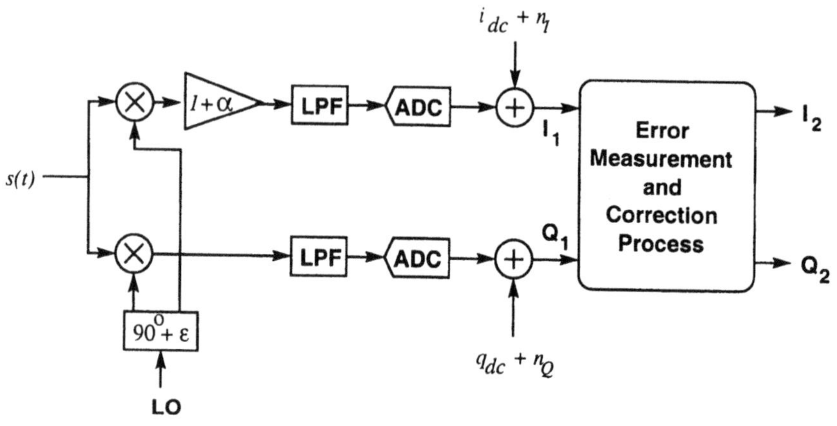

Figure 6.7: Quadrature receiver model used in the simulations.

Let the mismatch be $\alpha = 0.0506$, $\epsilon = 0.0873$. To calibrate the receiver, we apply a sinusoidal test signal with $A = 1$ and frequency located at $f_s/4$ above the LO. For this example, the output image power is shown as the solid curve with diamond in Figure 6.8, which is about -26 dB to the desired signal.

The statistical calibration method is then applied to correct the I/Q error. The step size of $\mu = 2^{-13}$ is used for the gain, phase and DC correction. Various SNR values are used in the simulation and the corresponding residual image power after the calibration are shown as the dashed curve with diamond in Figure 6.8. For comparison, the theoretical result of (6.30) is shown by the dotted curve. The simulation shows that the residual image power decreases by 20 dB when the SNR increases by 10 dB. For 20 dB SNR at the ADC output, we can have the residual image as low as -70 dB to the desired signal. The theoretical and simulated curves are well matched for RI within -90 dB. When the SNR is further increased, the residual image power remains unchanged.

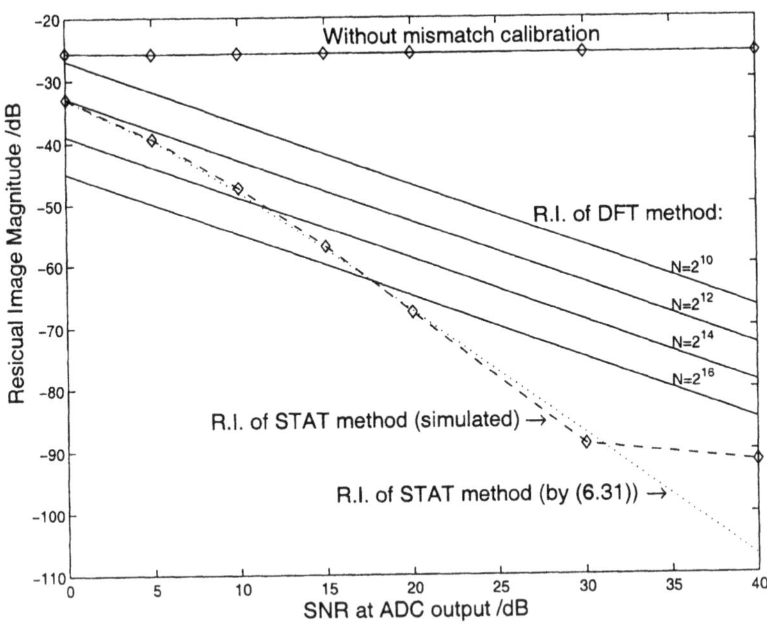

Figure 6.8: Residual image magnitude (relative to the desired signal magnitude) versus the SNR at ADC outputs.

This is because that the accuracies of the correction coefficients P and E are limited by the step size μ.

The calibration results by DFT method are also shown in Figure 6.8. For fixed data samples N used in the DFT calculation, the residual image power decreases by $10dB$ when the SNR increases by $10dB$. Doubling N one obtains 3 dB improvement on the residual image. For 20 dB SNR at the ADC output, and to have -70 dB of image rejection, the number of data samples required is larger than 70,000.

Figure 6.9 shows the convergency of P and E. The simulation conditions are: test signal amplitude $A = 1$, frequency $f_i = f_s/4$, SNR $= 30dB$, step size $\mu = 2^{-13}$ and mismatches of $\alpha = 0.0506$ and $\epsilon = 0.0873$. The final value of E and P are -0.052 and -0.0833 respectively, and the convergency time constants of E and P are 7600 and 14300 samples respectively. The result is in line with equation (6.32) and (6.31).

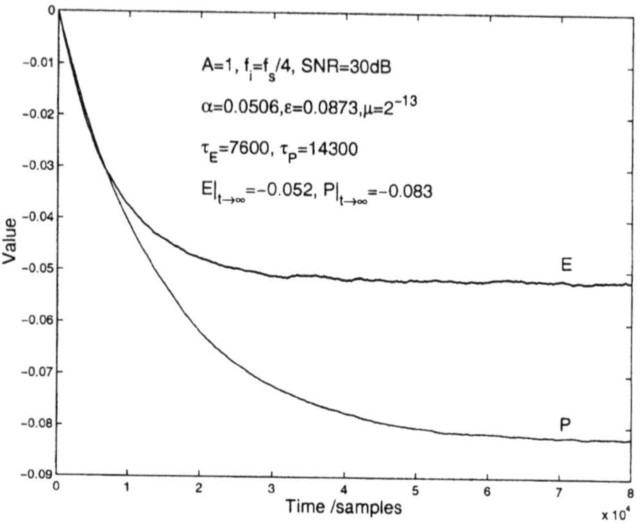

Figure 6.9: Convergency of P and E.

Lastly, the differences between the DFT and the statistical estimations of the correction coefficients are listed below.

- To calibrate frequency-independent mismatches, the DFT method needs sinusoidal test signal as the calibration source but the statistical method does not need it. The received signal itself can be used as the calibration signal.
- To calibrate the mismatches at different frequencies, the DFT method requires different length of DFT operations. But in the statistical method, the digital circuitry is the same for calibration at any frequencies. This becomes important in wideband calibration as presented in the next section.
- Noise effect. In the DFT method, there is no lower limit in the calibration accuracy due to noise. The noise effect can be reduced to any degree at the cost of the increasing in the number of samples used in the DFT calculation. In the statistical method, the SNR at the ADC outputs sets a lower bound of the maximum image rejection achievable.

To the authors' opinion, the statistical method is better in a general sense.

6.5 Wideband Calibration

As described in the last section, I/Q mismatches, or equivalently the correction parameters E and P, can be measured at any frequencies through the injection of sinusoidal test signals at the frequencies of interest. If we have these Es and Ps, then we can construct filters (denoted as $E(z)$ and $P(z)$) to correct the frequency-dependent mismatches [15]. Such system is particularly important for wideband receivers as their mismatches can be frequency-dependent in the band of interest.

Suppose we have measured the mismatches at M frequencies, $f_i, i = 1, 2, \cdots, M$, and the corresponding correction parameters are (E_i, P_i), $i = 1, 2, \ldots 1, M$. We want to construct two filters $E(z)$ and $P(z)$ whose values evaluated at f_i equal to E_i and P_i respectively. The desired filters $E(z)$ and $P(z)$ can be constructed by Lagrangian polynomials:

$$E(z) = \sum_{i=1}^{M} \frac{\prod_{k=1,k\neq i}^{M}(z + z^{-1} - 2\cos\omega_k)}{\prod_{k=1,k\neq i}^{M} 2(\cos\omega_i - \cos\omega_k)} E_i, \tag{6.36}$$

$$P(z) = \sum_{i=1}^{M} \frac{\prod_{k=1,k\neq i}^{M}(z + z^{-1} - 2\cos\omega_k)}{\prod_{k=1,k\neq i}^{M} 2(\cos\omega_i - \cos\omega_k)} P_i, \tag{6.37}$$

where $z = e^{j\omega}$, $\omega_i = 2\pi f_i/f_s$ and f_s is the sampling frequency. The numerator of (6.36) is determined to let the coefficient of E_i being equal to zero when (6.36) is evaluated at $z = e^{j\omega_k}, k \neq i$. The denominator of (6.36) is determined to let the coefficient of E_i being equal to one when (6.36) is evaluated at $z = e^{j\omega_i}$. The idea behind is to interpolate the values of (E, P) at frequencies other than f_i.

It can be easily verified that $E(z) = E_i$ and $P(z) = P_i$ at $f = f_i$, $i = 1, \ldots, M$. Besides, the impulse responses of $E(z)$ and $P(z)$ are finite with $2(M-1)$ tap, non-causal and symmetric. By substituting the numerical values of E_i, P_i and $\cos\omega_i$ to (6.36) and (6.37), we obtain:

$$E(z) = a_0 + a_1(z^1 + z^{-1}) + a_2(z^2 + z^{-2}) + \ldots + a_{M-1}(z^{M-1} + z^{-(M-1)}) \tag{6.38}$$

$$P(z) = b_0 + b_1(z^1 + z^{-1}) + b_2(z^2 + z^{-2}) + \ldots + b_{M-1}(z^{M-1} + z^{-(M-1)}) \tag{6.39}$$

where $a_0, a_1, \ldots, a_{M-1}$ and $b_0, b_1, \ldots, b_{M-1}$ are constants whose value can be estimated from (6.36) and (6.37).

Replacing the E and P block in Figure 6.3 with digital FIR filters $E(z)$ and $P(z)$ respectively results in the wideband I/Q mismatch correction system as shown in Figure 6.10(a). Note that $(M-1)^{th}$ delays are added on the I and Q paths due to the non-causality of $E(z)$ and $P(z)$.

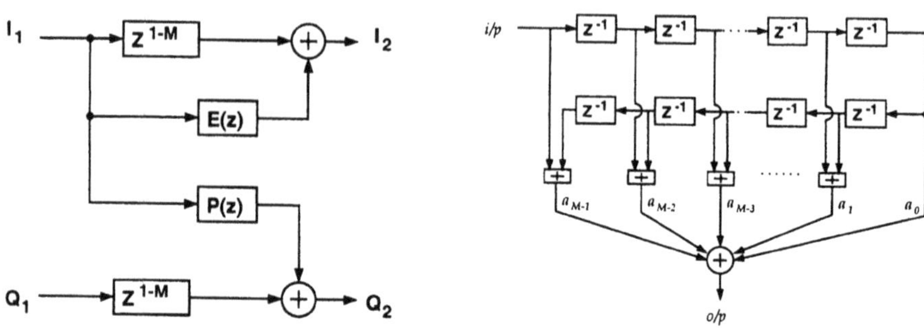

Figure 6.10: (a) The wideband mismatch correction system; (b) The filter structure of $E(z)$ and $P(z)$.

The filter architecture of $E(z)$ and $P(z)$ is shown in Figure 6.10(b). M additions and M multiplications are required in each filter. Two additions are required in Figure 6.10(a). So the real-time computing load of this system is $(4M + 2) \times f_s$ floating point operations (FLOPS) per second, where f_s is the output rate of the ADCs of the receiver.

The estimation of the coefficients of $E(z)$ and $P(z)$ involves trigonometric operations that increase digital circuit complexity. However, as these calculations are performed only once in the beginning of the calibration process, the computational load can be affordable.

6.5.1 Simulation and Experimental Results

High level simulations have been conducted to verify the wideband calibration system. The receiver model used in the simulation is same as that of Figure 6.7.

As an example, the receiver is assumed to have the channel mismatches linearly dependent on the frequency within the band of interest as shown in Figure 6.11. Although the mismatches are assigned arbitrarily, but such a linear model can be very close to real situations. The simulation is carried out with a point-by-point basis, i.e., the input test signal frequency is swept over the whole band of interest, and at each frequency corresponding values of α, ϵ are assigned to the receiver. The image magnitude in the receiver output is then measured at each frequency. The simulated image response of the receiver under such mismatch conditions is shown by the solid curve in Figure 6.12. The image rejection varies from $-25dB$ to $-30dB$ over the band of interest (from DC to $f_s/2$).

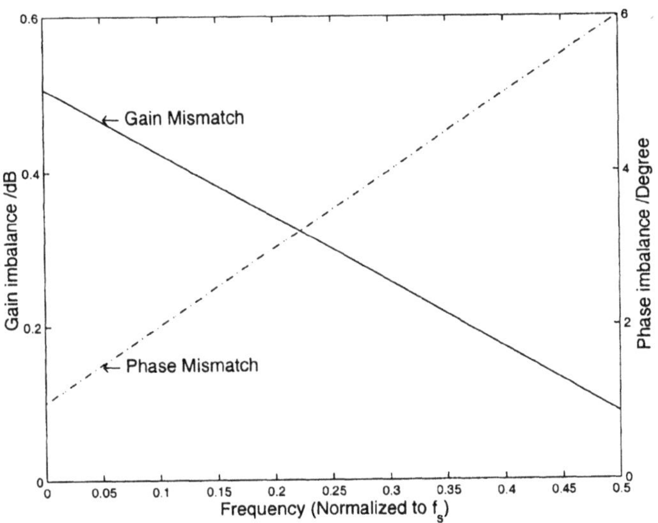

Figure 6.11: Receiver mismatch conditions.

The receiver is then calibrated at $f = 0.1f_s$, $0.2f_s$, $0.3f_s$ and $0.4f_s$. The statistical method was used to obtain the correction coefficients (E P) at these frequencies. The obtained values are listed in Table 6.1.

Table 6.1: Estimated E and P parameters in the example.

f	$0.1f_s$	$0.2f_s$	$0.3f_s$	$0.4f_s$
P	-0.03323	-0.05032	-0.0677	-0.08545
E	-0.04820	-0.03978	-0.0315	-0.02334

By (6.36) and (6.37), the two FIR filters $E(z)$ and $P(z)$ (see Figure 6.10(a)) are obtained as:

$$\begin{aligned} P(z) &= 9.2957 \times 10^{-4}(z^3 + z^{-3}) - 1.4256 \times 10^{-4}(z^2 + z^{-2}) \\ &\quad +0.016498(z^1 + z^{-1}) - 0.059254 \qquad (6.40) \\ E(z) &= -4.4197 \times 10^{-4}(z^3 + z^{-3}) - 5.9838 \times 10^{-5}(z^2 + z^{-2}) \\ &\quad -0.007856(z^1 + z^{-1}) - 0.03573 \qquad (6.41) \end{aligned}$$

After the correction, the image response is as shown by the dotted curve in Figure 6.12. Four notches are clearly observed at $f = 0.1f_s$,$0.2f_s$,$0.3f_s$ and

$0.4f_s$ which correspond to the calibration frequencies. It can be seen that the image are further suppressed by more than $25dB$ over the band of interest after the calibration.

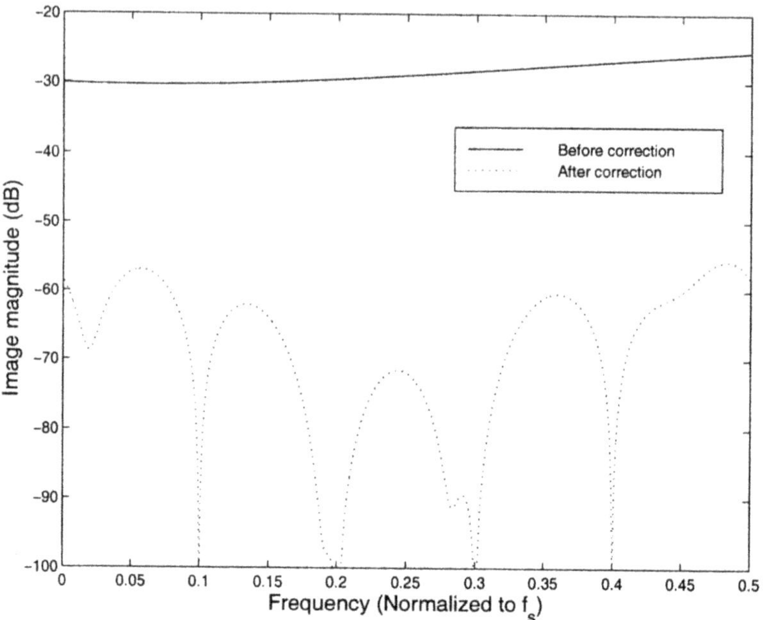

Figure 6.12: Image response before and after applying the wideband calibration method.

The wideband calibration method has been applied to an I/Q radio telescope receiver [16]. The receiver has an IF input of 200 to 220 MHz. The down converted I and Q baseband signals are digitised by 12 bit A/D converters with a 40 MHz sampling rate.

Gunst [16] reported that after the calibration, the achieved image rejection ratios averaged over the band of interest are 63 dB and 78 dB if 9 and 39 calibration frequencies are used respectively. Over a band of 20 MHz this corresponds to a performance improvement of 38 dB and 53 dB respectively.

6.6 Summary

Digital I/Q mismatch calibration methods have been discussed in this chapter to solve the self-image problem existing in direct conversion and complex-IF receivers. Existing methods developed for calibrating frequency-independent mismatch are inadequate for wideband applications.

A wideband calibration technique has been proposed to correct the frequency dependent I and Q channel imbalances. It involves an error measurement process and an error correction process, all implemented in baseband digital domain. Two error measurement methods, namely the DFT and statistical methods, are discussed and we conclude that the latter is more suitable for measuring mismatches at arbitrary frequency.

The real-time computational power of the wideband calibration system is $(4M + 2) \times f_s$ FLOPS per second, where M is the order of the system and f_s is the output rate of the ADCs of the receiver. As demonstrated by high-level simulations, the wideband calibration method can significantly improve the receiver's image rejection over the whole band of interest. The method has been successfully applied to a wideband radio telescope receiver.

References

[1] Behzad Razavi, *RF Microelectronics*, Prentice-Hall, 1998.

[2] E. van der Zwan, K. Philips, and C. Bastiaansen, "A 10.7MHz IF-to-basebad $\Delta\Sigma$ A/D conversion system for AM/FM radio receivers," in *Digest of Technical Papers, IEEE Int. Solid-State Circuit Conference*, Feb. 2000, pp. 340–341.

[3] Raf L. J. Roovers, "Wide-band A/D conversion for base stations," in *Circuits and Systems for Wireless Communications*, M. Helfenstein and G. S. Moschytz, Eds., pp. 187–196. Kluwer Academic Publishers, 2000.

[4] F.E. Churchill, G.W. Ogar, and B.J. Thompson, "The correction of I and Q errors in a coherent processor," *IEEE Transactions on Aerospace and Electronic Systems*, vol. AES-17, no. 1, pp. 131–137, Jan 1981.

[5] M.D. MacLeod, "Fast calibration of IQ digitiser systems," *Electronic Engineering, GB, Morgan-Grampian Ltd., London*, vol. 62, no. 757, pp. 41–43, 1990.

[6] R. A. Green, R. Anderson-Sprecher, and J. W. Pierre, "Quadrature receiver mismatch calibration," *IEEE Tran. Signal Processing*, vol. 47, no. 11, pp. 3130–3133, Nov. 1999.

[7] M.D. Kulkarni and Alexander B. Kostinski, "A simple formula for monitoring quadrature phase error with arbitrary signals," *IEEE Tran. Geoscience and remote sensing*, vol. 33, no. 3, pp. 799–802, May 1995.

[8] Daihong Fu and K. C. Dyer, "A digital background calibration technique for time-interleaved analog-to-digital converters," *IEEE J. of Solid-state circuits*, vol. 33, no. 12, pp. 1904–1911, Dec 1998.

[9] T.H. Shu, B.S. Song, and K. Bacrania, "A 13-b 10-Msamples ADC digitally calibrated with oversampling delta-sigma converter," *IEEE J. of Solid-state circuits*, vol. 30, no. 4, pp. 443–452, Apr 1995.

[10] A.N. Karanicolas, H.S. Lee, and K. L. Bacrania, "A 15-b 1-Msample/s digitally self-calibrated pipeline ADC," *IEEE J. Solid-state circuits*, vol. 28, no. 12, pp. 1207–1215, Dec 1993.

[11] J.P.Y. Lee, "Wideband I/Q demodulators: measurement technique and matching characteristics," *IEE Proc.-Radar, Sonar and Navigation*, vol. 143, no. 5, pp. 300–306, October 1996.

[12] K.P. Pun, J.E. Franca, and C. Azeredo Leme, "A digital method for the correction of I/Q phase errors in complex sub-sampling mixers," in *Proc. IEEE Southwest Symposium on Mixed Signal Design*, San Diego, California, USA, Feb. 2000, pp. 171–174.

[13] K.P. Pun, J.E. Franca, and C. Azeredo Leme, "Dynamic I/Q mismatch calibration in quadrature receivers," in *Abstract Book, 1st Portugal-China Workshop on Solid-State Circuits*, Shanghai, China, Oct. 2000, pp. 13–14.

[14] The Math Works Inc., *User's guide, MATLAB 5*, 1996.

[15] K.P. Pun, J.E. Franca, and C. Azeredo Leme, "Wideband digital correction of I and Q mismatch in quadrature radio receivers," in *Proc. IEEE Int. Symposium on Circuits and Systems*, Geneva, Switzerland, May 2000, vol. V, pp. 661–664.

[16] A.W. Gunst and G. W. Kant, "Application of digital wideband mismatch calibration to an i/q receiver," in *Proc. Internation Symposium on Circuits and Systems, Arizona*, May 2002, vol. 3, pp. 484 –487.

Appendix 6.A

This appendix proofs (6.4) and (6.5) for the estimation of E and P from the DFT on the outputs of a test signal.

Since the test signal output frequency is at 1/4 of the sampling frequency, the DFT components $\tilde{s}_1[1]$ and $\tilde{s}_1[3]$ are the desired and image signals respectively. By the linear property of DFT, we have

$$\begin{aligned} s_1[1] &= \tilde{I}_1[1] + j\tilde{Q}_1[1], \\ s_1[3] &= \tilde{I}_1[3] + j\tilde{Q}_1[3], \\ s_1^*[1] &= \tilde{I}_1^*[1] - j\tilde{Q}_1^*[1] \\ &= \tilde{I}_1[3] - j\tilde{Q}_1[3], \qquad (6.42) \\ \tilde{I}_1[3] &= \frac{1}{2}(\tilde{s}_1^*[1] + \tilde{s}_1[3]), \qquad (6.43) \end{aligned}$$

where * denotes the complex conjugate. Equation (6.42) uses the property that the DFT of real signals (I_1 and Q_1 are real signals) are conjugate symmetric.

From (6.3), a complex component $(E + jP)I_1$ is added to s_1 so that the resultant signal $s_2 = s_1 + (E+jP)I_1$ contains no image component, i.e., $\tilde{s}_2[3] = 0$. Therefore we must have

$$\tilde{s}_1[3] + (E + jP)\tilde{I}[3] = 0. \qquad (6.44)$$

From (6.43) and (6.44), we obtain

$$E + jP = -\frac{2\tilde{s}_1[3]}{\tilde{s}_1^*[1] + \tilde{s}_1[3]}. \qquad (6.45)$$

Therefore, the correction coefficients E and P are obtained:

$$E = -Re\{\frac{2\tilde{s}_1[3]}{\tilde{s}_1^*[1]+\tilde{s}_1[3]}\} \qquad (6.46)$$

$$P = -Im\{\frac{2\tilde{s}_1[3]}{\tilde{s}_1^*[1]+\tilde{s}_1[3]}\}. \qquad (6.47)$$

Appendix 6.B

The statistical estimation of parameters P and E as shown in Figure 6.5 and Figure 6.6 respectively, as well as their residual errors due to noise are derived in this appendix.

Let the I and Q inputs of the phase error estimation system be

$$\begin{aligned} I_1 &= I_1' + n_I, \quad \text{and} \qquad (6.48) \\ Q_1 &= Q_1' + n_Q, \qquad (6.49) \end{aligned}$$

respectively, where

$$I_1' = (1+\alpha)A\cos(\omega_i t), \tag{6.50}$$
$$Q_1' = A\sin(\omega_i t + \epsilon), \tag{6.51}$$

and n_I and n_Q are zero-mean noise of I and Q channels with variances of

$$< n_I^2 > = (1+\alpha)^2\sigma^2, \tag{6.52}$$
$$< n_Q^2 > = \sigma^2, \tag{6.53}$$

respectively, where $< \bullet >$ represents the time averaging. Note that the signal-to-noise ratios at the ADC outputs of I and Q channels are assumed equal. The phase error correction system of Figure 6.5 results in $< I_{out}Q_{out} >= 0$, i.e.,

$$< (I_1' + n_I)(Q_1' + n_Q + P(I_1' + n_I)) >= 0, \tag{6.54}$$
$$< I_1'Q_1' > + < n_Q I_1' > +P < I_1'^2 > + < n_I Q_1' > + < n_I n_Q > +2P < n_I I_1' > +P < n_I^2 >= 0, \tag{6.55}$$
$$< I_1'Q_1' > +P < I_1'^2 >= -P < n_I^2 > . \tag{6.56}$$

The last step in the above derivation have used the assumption that noises n_I and n_Q are uncorrelated to each other and to I_1' and Q_1'. Since

$$< I_1'Q_1' > = \frac{1}{2}(1+\alpha)A^2\sin\epsilon, \tag{6.57}$$
$$< I_1'^2 > = \frac{1}{2}A^2(1+\alpha)^2, \tag{6.58}$$

and from (6.52), (6.56) we obtain

$$P = -\frac{\sin\epsilon}{1+\alpha}\frac{\text{SNR}}{1+\text{SNR}}, \tag{6.59}$$

where $\text{SNR} \equiv \frac{<I_1'^2>}{<n_I^2>} = A^2/2\sigma^2$ is the signal-to-noise ratio at the ADC outputs.

After the phase correction, the I channel signal keeps unchanged, while the Q channel signal becomes

$$Q_2 = Q_1 + PI_1 = Q_1' + PI_1' + (n_Q + Pn_I). \tag{6.60}$$

Denoting $Q_2' = Q_1' + PI_1'$, from (6.59), we have

$$\begin{aligned} Q_2' &= Q_1' + PI_1' \\ &= A\sin(\omega_i t + \epsilon) - \frac{\sin\epsilon}{1+\alpha}\frac{\text{SNR}}{1+\text{SNR}}A(1+\alpha)\cos(\omega_i t) \\ &= A\cos\epsilon\sin(\omega_i t) + \frac{1}{1+\text{SNR}}A\sin\epsilon\cos(\omega_i t) \\ &= A_{q2}\sin(\omega_i t + \epsilon'), \end{aligned} \tag{6.61}$$

where the amplitude A_{q2} and residual phase error ϵ' are

$$A_{q2} = A\sqrt{\cos^2\epsilon + \frac{\sin^2\epsilon}{(1+\text{SNR})^2}}, \tag{6.62}$$

$$\begin{aligned}\epsilon' &= \arctan\left(\frac{\tan\epsilon}{1+\text{SNR}}\right) \\ &\approx \frac{\tan\epsilon}{1+\text{SNR}} \quad (\text{Since } \epsilon' \ll 1).\end{aligned} \tag{6.63}$$

The I and Q data after the phase correction, (6.48) and (6.60) respectively, are then applied to the gain correction system shown in Figure 6.6. The system results in $< I_{out}^2 - Q_{out}^2 >= 0$, i.e,

$$< (1+E)^2(I_1' + n_I)^2 - (Q_2' + n_Q + Pn_I)^2 >= 0, \tag{6.64}$$

$$(1+E)^2(< I_1'^2 > + < n_I^2 >) - (< Q_2'^2 > + < n_Q^2 > + P^2 < n_I^2 >) = 0. \tag{6.65}$$

The last equation above uses the fact that n_I and n_Q are uncorrelated to each other and to I_1' as well as Q_2'. Substituting (6.52), (6.53), (6.59), (6.58) and (6.61) to (6.65), we obtain

$$\begin{aligned}(1+E)^2\left(\frac{1}{2}A^2(1+\alpha)^2 + (1+\alpha)^2\sigma^2\right) &= \frac{1}{2}A^2\left(\cos^2\epsilon + \frac{\sin^2\epsilon}{(1+\text{SNR})^2}\right) \\ &\quad +\sigma^2 + \frac{\text{SNR}^2\sin^2\epsilon}{(1+\text{SNR})^2}\sigma^2, \\ (1+E)^2(1+\alpha)^2(1+\text{SNR}) &= \text{SNR}\left(\cos^2\epsilon + \frac{\sin^2\epsilon}{(1+\text{SNR})^2}\right) \\ &\quad +1 + \frac{\text{SNR}^2\sin^2\epsilon}{(1+\text{SNR})^2}, \\ (1+E)^2 &= \frac{\cos^2\epsilon}{(1+\alpha)^2}\left(1+\tan^2\epsilon\frac{1+2\text{SNR}}{(1+\text{SNR})^2}\right).\end{aligned}$$

Therefore,

$$E = \frac{\cos\epsilon}{1+\alpha}\sqrt{1+\tan^2\epsilon\frac{1+2\text{SNR}}{(1+\text{SNR})^2}} - 1. \tag{6.66}$$

After the gain correction, the Q channel signal (6.60) keeps unchanged, while the I channel signal becomes $I_2 = (1+E)I_1' + (1+E)n_I$. Denote $I_2' = (1+E)I_1'$. From (6.66),

$$I_2' = \left(A\cos\epsilon\sqrt{1+\tan^2\epsilon\frac{1+2\text{SNR}}{(1+\text{SNR})^2}}\right)\cos(\omega_i t). \tag{6.67}$$

From (6.61) and (6.67), we have

$$\begin{aligned}\frac{< I_2'^2 >}{< Q_2'^2 >} &= \frac{\cos^2\epsilon + (\sin^2\epsilon)(1+2\mathrm{SNR})/(1+\mathrm{SNR})^2}{\cos^2\epsilon + (\sin^2\epsilon)/(1+\mathrm{SNR})^2} \\ &= 1 + \frac{(\sin^2\epsilon)2\mathrm{SNR}/(1+\mathrm{SNR})^2}{\cos^2\epsilon + (\sin^2\epsilon)/(1+\mathrm{SNR})^2} \\ &\approx 1 + \tan^2\epsilon\frac{2\mathrm{SNR}}{(1+\mathrm{SNR})^2} \quad \text{for SNR} \gg 1 \\ &\approx 1 + \frac{2\tan^2\epsilon}{\mathrm{SNR}}. \end{aligned} \tag{6.68}$$

Therefore, the residual gain error α' is given by

$$\begin{aligned}\alpha' &= 1 - \frac{< |I_2'| >}{< |Q_2'| >}, \\ &\approx \frac{\tan^2\epsilon}{\mathrm{SNR}} \quad \text{for SNR} \gg 1. \end{aligned} \tag{6.69}$$

Appendix 6.C

The convergence speeds of the error correction systems are analysed in this appendix.

First, consider the phase correction system of Figure 6.5. The input $x[i]$ to the accumulator is

$$\begin{aligned}x[i] &= I_{in}[i]Q_{out}[i] \\ &= I_{in}[i](Q_{in}[i] + P[i]I_{in}[i]) \\ &= I_{in}[i]Q_{in}[i] + P[i]I_{in}^2[i].\end{aligned}$$

Let the I and Q inputs be:

$$I_{in}[i] = A(1+\alpha)\cos(2\pi i/T),$$

$$Q_{in}[i] = A\sin(2\pi i/T + \epsilon),$$

where T is the period of I_{in} and Q_{in}, then the output $P[k]$ of the accumulator can be interpreted as:

$$\begin{aligned}
-P[k] &= \mu_P \sum_{i=0}^{k} x[i] \\
&= \mu_P \sum_{i=0}^{k/T} \sum_{j=(i-1)T+1}^{iT} x[j] \quad \text{for } k = 0, T, 2T, 3T, \ldots \\
&= \mu_P \sum_{i=0}^{k/N} \sum_{j=(i-1)T+1}^{iT} (I_{in}[j]Q_{in}[j] + P[j]I_{in}^2[j]) \\
&\approx \mu_P \sum_{i=0}^{k/T} \sum_{j=(i-1)T+1}^{iT} (I_{in}[j]Q_{in}[j] + P[iT]I_{in}^2[j]) \\
&= \mu_P \sum_{i=0}^{k/T} \sum_{j=(i-1)T+1}^{iT} (A^2(1+\alpha)\cos(2\pi j/T)\sin(2\pi j/T + \epsilon) \\
&\qquad\qquad + P[iT]A^2(1+\alpha)^2\cos^2(2\pi j/T)) \\
&= \mu_P \sum_{i=0}^{k/T} \left(\frac{A^2(1+\alpha)T\sin\epsilon}{2} + P[iT]\frac{A^2(1+\alpha)^2 T}{2} \right) \\
&= \frac{\mu_P A^2(1+\alpha)T}{2} \sum_{i=0}^{k/T} (\sin\epsilon + P[iT](1+\alpha)), \quad k = 0, T, 2T, 3T, \ldots.
\end{aligned}$$

In the above equation, $P[j]$ is assumed to have a very small variation over every period, i.e., $P[j] \approx P[iT]$ for $(i-1)T + 1 < j < iT$. Let

$$P'[k/T] = P[k] \quad \text{for } k = 0, T, 2T, \ldots \tag{6.70}$$

we have

$$-P'[m] = \mu_P \frac{A^2(1+\alpha)T}{2} \sum_{i=0}^{m} (\sin\epsilon + P'[i](1+\alpha)) \quad m = 0, 1, 2, 3, \ldots \tag{6.71}$$

From the above equation, we can construct a linear system as shown in Figure 6.13.

The z-domain solution of P' is:

$$-P'(z) = \frac{\mu_P A^2(1+\alpha)T}{2} \frac{1}{1-z^{-1}} \left(\frac{\sin\epsilon}{1-z^{-1}} + P(1+\alpha) \right), \tag{6.72}$$

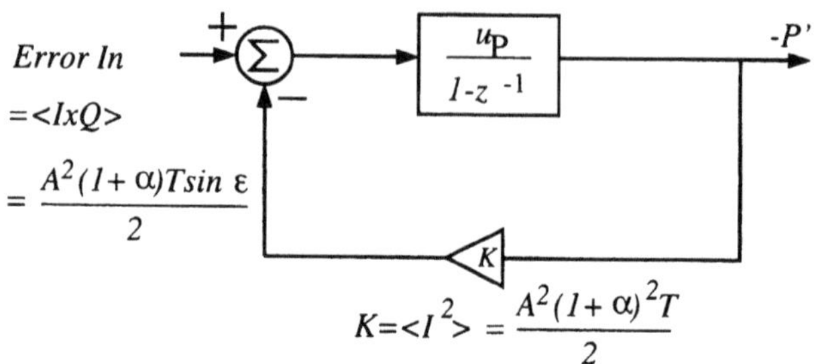

Figure 6.13: A linear model for the phase error correction.

$$\therefore P'(z) = \frac{\sin\epsilon}{1+\alpha}\left(\frac{1}{1-z^{-1}} - \frac{1}{1+\mu_P A^2(1+\alpha)^2 T/2 - z^{-1}}\right). \tag{6.73}$$

Applying inverse z-transform, we have

$$P'[m] = \frac{\sin\epsilon}{1+\alpha}\left(1 - \left(1 + \frac{\mu_P A^2(1+\alpha)^2 T}{2}\right)^{-(m+1)}\right) u[m] \tag{6.74}$$

where $u[m]$ is the step function. Since $\mu_P A^2(1+\alpha)^2 T/2 > 0$, the final value of P' is

$$P'|_{t\to\infty} = \frac{\sin\epsilon}{1+\alpha}. \tag{6.75}$$

If μ_P is chosen small enough so that $\mu_P A^2(1+\alpha)^2 T/2 \ll 1$, it can be easily shown that the convergence time constant of P' is:

$$\tau_{P'} = \frac{2}{\mu_P A^2(1+\alpha)^2 T}, \tag{6.76}$$

where $\tau_{P'}$ is the time (in terms of the number of samples) needed for P' to reach $(1-e^{-1})$ of its final value. From (6.70), we have the convergence time constant of P:

$$\tau_P = \frac{2}{\mu_P A^2(1+\alpha)^2} \cong \frac{2}{\mu_P A^2}. \tag{6.77}$$

Similarly, we can get the convergence time constant of E as:

$$\tau_E = \frac{\pi}{2\mu_E A(1+\alpha)} \cong \frac{\pi}{2\mu_E A}. \tag{6.78}$$

For the DC error correction circuit of Figure 6.4, the convergence time constant is $\tau_{dc} = \frac{1}{\mu_{dc}}$.

Chapter 7

Blind Compensation of I/Q Mismatches

7.1 Introduction

As pointed out in Chapter six, the gain mismatch and phase error between the I and Q channels in a quadrature receiver, adversely affect the performance of the receiver, by generating image components of the receiving signal. Its effect is very problematic in a receiver designed for multi-channel reception.

Existing estimation and calibration methods for I/Q mismatches [1] [1, 2, 3, 4, 5], all assume that I/Q mismatches are frequency-independent. In Chapter six, a frequency-dependent I/Q mismatch calibration method was proposed. It requires the injection of sinusoidal test tones in order to estimate the mismatch information. Therefore, it has the disadvantage of needing an extra oscillator, either off-chip or on-chip. Besides, once the calibration process is done, the correction program can not adapt to time variations of I/Q mismatches.

In this chapter, a digital system based on an adaptive signal separation algorithm is constructed to correct the frequency-dependent I/Q mismatches in a multi-channel quadrature receiver [6]. By using the complex conjugate $(I-jQ)$ of the receiver's output as the reference, the proposed method can effectively separate the image interferer and the desired signal that are mixed together in the receiver's output, and is therefore referred to as a *signal-image separation* method. Comparing to the calibration method presented in Chapter six, this system requires no test signal injections and therefore can be implemented

[1] For simplicity the term "I/Q mismatches", or "I/Q imbalances" is used in this chapter to signify both the gain mismatch and phase error.

much more easily. Another advantage is that it can adapt to the possible time-variation of I/Q mismatches.

This chapter is arranged as follows. Firstly, the concept of signal-image separation is introduced. Secondly, the adaptive signal separation algorithm used in this method is presented. Thirdly, verifications of the proposed method by high-level simulations are given. Lastly, the chapter is summarised.

7.2 The Concept of Signal-Image Separation

Suppose in the absence of I/Q mismatches, the I and Q outputs of a receiver are $x_{1,I}(k)$ and $x_{1,Q}(k)$ respectively. Denote their z-transforms as $X_{1,I}(z)$ and $X_{1,Q}(z)$, where $z = e^{j\omega}$ and ω is the angular frequency.

Now, consider the presence of frequency-dependent gain and phase imbalances. Assume the I and Q paths of the receiver have mismatched gains of $A_I(\omega)$ and $A_Q(\omega)$ respectively, and phase errors of $\theta_I(\omega)$ and $\theta_Q(\omega)$ respectively. Then the complex output $Y_1(e^{j\omega})$ of the receiver can be expressed as:

$$\begin{aligned} Y_1(e^{j\omega}) &= Y_{1,I}(e^{j\omega}) + jY_{1,Q}(e^{j\omega}) \\ &= A_I(\omega)e^{j\theta_I(\omega)}X_{1,I}(e^{j\omega}) + jA_Q(\omega)e^{j\theta_Q(\omega)}X_{1,Q}(e^{j\omega}) \\ &= H_{cm}(\omega)X_1(e^{j\omega}) + H_{dif}(\omega)X_1^*(e^{-j\omega}), \end{aligned}$$

where $X_1(e^{j\omega}) = X_{1,I}(e^{j\omega}) + jX_{1,Q}(e^{j\omega})$, $X_1^*(e^{-j\omega}) = X_{1,I}(e^{j\omega}) - jX_{1,Q}(e^{j\omega})$ is the $z-$transform of the image component $x_1^* = x_{1,I} - jx_{1,Q}$, H_{cm} and H_{dif} are respectively the common-mode and differential responses of I and Q channels, given by:

$$H_{cm}(\omega) = \left[A_I(\omega)e^{j\theta_I(\omega)} + A_Q(\omega)e^{j\theta_Q(\omega)}\right]/2 \tag{7.1}$$

$$H_{dif}(\omega) = \left[A_I(\omega)e^{j\theta_I(\omega)} - A_Q(\omega)e^{j\theta_Q(\omega)}\right]/2. \tag{7.2}$$

The mirror image $X_1^*(e^{-j\omega})$ is coupled to the output through the term $H_{dif}(\omega)$ which represents the channel mismatches and is in general unknown and possibly time-varying. Fig. 7.1 shows this effect. For perfect channel matching, we have $A_I(\omega) = A_Q(\omega)$ and $\theta_I(\omega) = \theta_Q(\omega)$ for all ω. The term $H_{dif}(\omega)$ becomes zero and the output Y_1 will be "clean", i.e., containing no image components.

If we have a reference signal for image, then classical adaptive noise canceller, for example, the well-known Widrow's least mean squares (LMS) scheme [7], can be used to cancel the image interference. The system model of such an adaptive noise canceller is shown in Figure 7.2. The receiver output y_1 is used as the primary input. For the reference input y_2, an ideal source is the image x_1^*

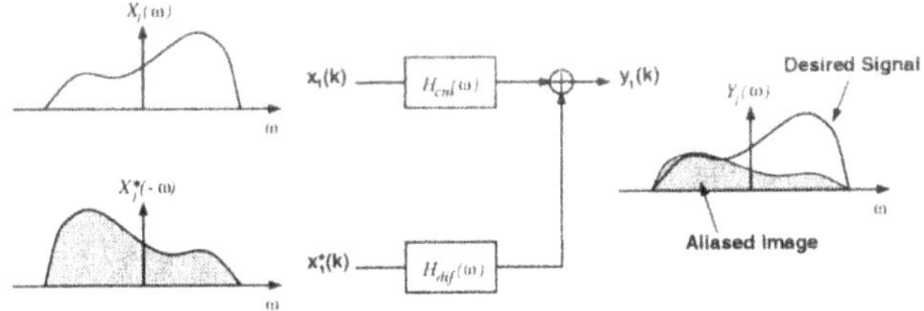

Figure 7.1: Modelling of the I/Q mismatch effect.

itself $(x_{1,I} - jx_{1,Q})$. (Note that all the signals and the adaptive filter $w_1(z)$ in Figure 7.2 should be in complex form in this case.) The primary input y_1 and reference input x_1^* are correlated in an unknown way $H_{dif}(\omega)$. The principle of this noise canceller is that the $H_{dif}(\omega)$ can be identified by minimising the average power of the reconstructed, or estimated, signal and use it for cancelling the image x_1^* at the primary input. Mathematically, minimising the average power corresponds to identifying, or estimating, the unknown function $H_{dif}(\omega)$ by a least-squares fit of the reference input to the primary input.

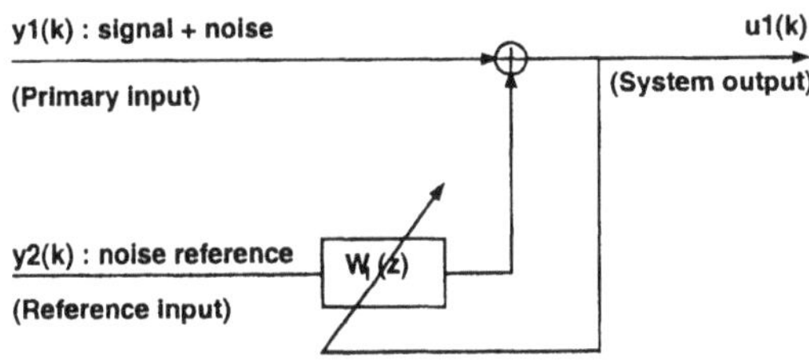

Figure 7.2: Classical adaptive noise canceller.

The LMS method can be very effective if one has the ideal reference x_1^*. However, to obtain x_1^* is just as difficult as to obtain the incorrupted desired signal x_1. A natural source for the reference input is the complex conjugate of the receiver output, i.e., y_1^*. However, the signal leakage problem is then arising. Note that the reference y_1^* contains a leakage from the desired signal x_1. It was proven that the signal-to-noise ratio at the output of Widrow's system is at best the noise-to-signal ratio at the reference input [7]. This phenomenon is also referred as *power inversion* [8]. As the noise-to-signal, or image-to-signal, ratio of y_1^* is exactly equal to the signal-to-noise, or signal-to-image, ratio of y_1, there is no point to add this noise canceller to the receiver.

Many methods have been proposed to deal with the signal leakage problem [9, 10, 11, 12, 13, 14]. In this thesis, a symmetric adaptive decorrelation (SAD) signal separation method, proposed by Gerven *et al* [15], is chosen.

The architecture of the SAD signal separation system is shown in Figure 7.3. The underlying principle is straightforward. It can be shown that the least squares criterion in the classical method is equivalent to the decorrelation of the signal estimate u_1 at the output with the noise reference, or image reference in this application. The image reference y_1^* contains parts of the desired signal x_1. Such a criterion makes little sense, and it would be better if a signal free image estimate were available. To obtain such a signal free image reference, a symmetric filter is added as shown in Figure 7.3. Decorrelation is now done between an image free signal estimate u_1 and a signal free image estimate u_2. The least squares criterion is replaced by the decorrelation criterion and due to its complete symmetry, the algorithm is a signal-image separator rather than an image canceller. Details of the algorithm is presented in the next section.

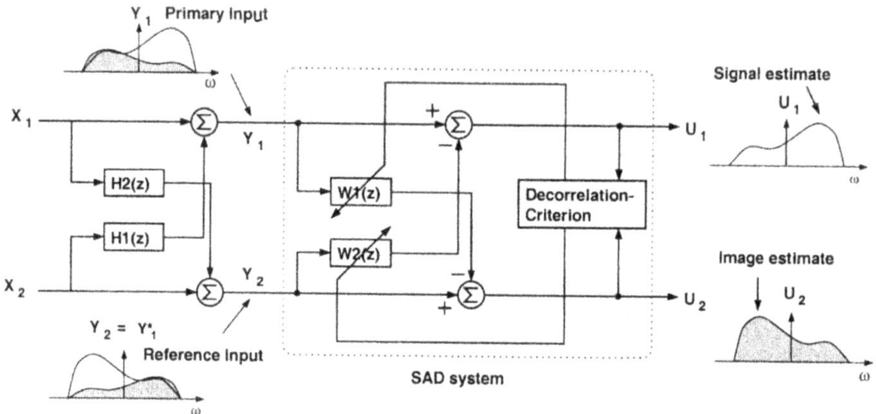

Figure 7.3: The symmetric adaptive decorrelation system.

7.3 Symmetric Adaptive Decorrelation Algorithm

First, let us look at the SAD algorithm for real signal processing as considered in [15].

The primary input $y_1(k)$ and the reference signal $y_2(k)$, in the presence of inter-coupling, are written as

$$y_1(k) = x_1(k) + \boldsymbol{h}_1(k) \otimes \boldsymbol{x}_2(k) \quad (7.3)$$
$$y_2(k) = x_2(k) + \boldsymbol{h}_2(k) \otimes \boldsymbol{x}_1(k), \quad (7.4)$$

where $\boldsymbol{x}_1(k)$ and $\boldsymbol{x}_2(k)$ are vectors of the uncorrelated signals to be estimated, $\otimes$ represents convolution. The $\boldsymbol{h}_1(k)$ and $\boldsymbol{h}_2(k)$ are the unknown cross-coupling functions to be identified.

The aim for the SAD signal separation system is to obtain signal estimates $u_1(k)$ and $u_2(k)$ by adaptive filtering of $y_1(k)$ and $y_2(k)$. The signal estimates $u_1(k)$ and $u_2(k)$ can be written as

$$u_1(k) = y_1(k) - \boldsymbol{w}_1^T(k) \otimes \boldsymbol{y}_2(k) \quad (7.5)$$
$$u_2(k) = y_2(k) - \boldsymbol{w}_2^T(k) \otimes \boldsymbol{y}_1(k), \quad (7.6)$$

where $\boldsymbol{w}_1(k)$ and $\boldsymbol{w}_1(k)$ are the coefficients of the adaptive filters, expressed as

$$\boldsymbol{w}_1(k) = [w_1^{(k)}(0)\ w_1^{(k)}(1)\ \ldots w_1^{(k)}(L_1 - 1)]^T \quad (7.7)$$
$$\boldsymbol{w}_2(k) = [w_2^{(k)}(0)\ w_2^{(k)}(1)\ \ldots w_2^{(k)}(L_2 - 1)]^T, \quad (7.8)$$

and L_1 and L_2 are the orders of the filters, and $\boldsymbol{y}_1(k)$ and $\boldsymbol{y}_2(k)$ are input signal vectors expressed by

$$\boldsymbol{y}_1(k) = [y_1(k)\ y_1(k-1)\ \ldots\ y_1(k - L_1 + 1)]^T \quad (7.9)$$
$$\boldsymbol{y}_2(k) = [y_2(k)\ y_2(k-1)\ \ldots\ y_2(k - L_2 + 1)]^T. \quad (7.10)$$

Filter coefficients are updated by the following formulae:

$$w_1^{(k+1)}(m) = w_1^{(k)}(m) + \mu_1 u_1(k) u_2(k-m), \quad m = 0, \cdots, (L_1 - 1) \quad (7.11)$$
$$w_2^{(k+1)}(n) = w_2^{(k)}(n) + \mu_2 u_2(k) u_1(k-n), \quad n = 0, \cdots, (L_2 - 1), \quad (7.12)$$

where μ_1 and μ_2 are the step-sizes that control the speed and stability of system ($0 < \mu_i < 2/\sigma_2^2$, where σ_2 is the variation of x_2). Numerically, the algorithm corresponds to the simplified Newton-Raphson zero search in the cross-correlation of u_1 and u_2, with the expected values replaced by their instantaneous sample estimates.

7.3.1 Complex Symmetry Adaptive Decorrelation Algorithm

The symmetry adaptive decorrelation algorithm proposed in [15] was for real signal processing. It was modified for complex signal processing by Yu *et al* in [16]. For complex signal processing, the adaptive filters $w_1(z)$ and $w_2(z)$ of the the signal separation system shown in Figure 7.3 must be in complex form. A complex filter ($W(z)$) can be realized by two real-coefficient sub-filters ($W_{Re}(z)$ and $W_{Im}(z)$) [17] as shown in Figure 7.4.

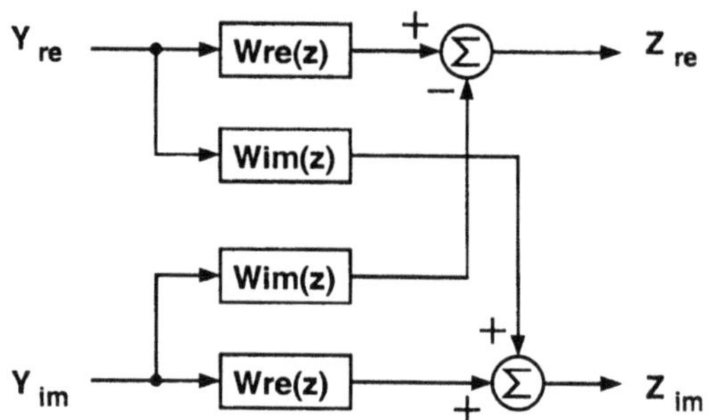

Figure 7.4: The complex filter structure.

The primary input $y_1(k)$ and the reference input $y_2(k)$ of Figure 7.3, in the presence of cross-coupling, can be expressed as

$$y_i(k) = x_i(k) + \boldsymbol{h}_i(k) \otimes \boldsymbol{x}_j(k) \quad i = 1, 2; \quad j = 2, 1, \tag{7.13}$$

where $\boldsymbol{x}_i(k)$ are vectors of the original uncorrelated signals to be estimated, and $\boldsymbol{h}_j(k)$ are vectors of the unknown cross-coupling functions to be identified. In the image cancellation application, $x_1(k)$ is the desired signal; $x_2(k)$ is its image ($x_2(k) = x_1^*(k)$); $y_2(k) = y_1^*(k)$; $h_1(k)$ is the I/Q mismatch term $h_{dif}(k)$ of (7.2); and $h_2(k) = h_{dif}^*(k)$.

The complex outputs $u_i(k) = u_{i,Re}(k) + ju_{i,Im}(k)$ of the signal separation system can be expressed as

$$u_i(k) = y_i(k) - \boldsymbol{w}_i^T(k) \otimes \boldsymbol{y}_j(k), \quad i = 1, 2, \quad j = 2, 1. \tag{7.14}$$

The cross-correlation $C_{xy}(m)$ between two complex signals $x(k)$ and $y(k)$ equals $E[x(k)y^*(k-m)]$, where E represents the expected value. Therefore, the coefficient updating equations (7.11)(7.12) are modified to

$$\boldsymbol{w}_i(k+1) = \boldsymbol{w}_i(k) + 2\mu_i u_i(k)\boldsymbol{u}_j^*(k), \quad i = 1, 2, \quad j = 2, 1, \tag{7.15}$$

where μ_i are the step-sizes which control the speed and stability of the SAD system, and can be chosen as integer powers of two so that the multiplication can be realized just by bit-shifting. The coefficients of the adaptive filters are

$$\boldsymbol{w}_i(k) = \boldsymbol{w}_{i,Re}(k) + j\boldsymbol{w}_{i,Im}(k), \quad i = 1, 2 \tag{7.16}$$

where

$$w_{i,Re}(k) = [w_{i,Re}^{(k)}(0)\ w_{i,Re}^{(k)}(1)\ \cdots\ w_{i,Re}^{(k)}(L_i - 1)]^T \tag{7.17}$$
$$w_{i,Im}(k) = [w_{i,Im}^{(k)}(0)\ w_{i,Im}^{(k)}(1)\ \cdots\ w_{i,Im}^{(k)}(L_i - 1)]^T, \tag{7.18}$$

and L_i are the order of the adaptive filters. Numerically, the (7.15) corresponds to the simplified Newton-Raphson zero search in the cross-correlation of complex variables u_1 and u_2, with the expected values replaced by their instantaneous sample estimates. In the image cancellation application, since $y_2(k) = y_1^*(k)$, by symmetry we will have $w_1(k) = w_2^*(k)$ if $L_1 = L_2$ and $\mu_1 = \mu_2$. If equation (7.15) converges, $u_1(k)$ and $u_2(k)$ will be decorrelated over the span of both filters:

$$C_{u_1 u_2}(m) = E[u_1(k)u_2^*(k-m)] = 0, \quad m = 0 \ldots L_1 \tag{7.19}$$
$$C_{u_2 u_1}(n) = E[u_2(k)u_1^*(k-n)] = 0, \quad m = 0 \ldots L_2. \tag{7.20}$$

7.3.2 Condition, convergence and noise

It is important to notice that a precondition of this algorithm is that the original signals $x_1(k)$ and $x_2(k)$ must be uncorrelated with each other. In the image cancellation application, $x_1(k)$ is the desired signal and $x_2(k)$ is the image of $x_1(k)$, i.e, $x_1^*(k)$. It can be easily proven that the condition corresponds to

$$C_{x_{1,I}x_{1,I}}(m) = C_{x_{1,Q}x_{1,Q}}(m) \quad \text{and} \tag{7.21}$$
$$C_{x_{1,I}x_{1,Q}}(m) = 0, \quad m = 0, 1, \cdots, max\{L_1, L_2\}. \tag{7.22}$$

Equation (7.21) means that the desired I and Q signals must have the same self-correlation. Equation (7.22) means that the desired I and Q signals must be uncorrelated with each other. In a communication system, such requirements are usually satisfied.

Convergence to the desired solution is not guaranteed. However, as suggested in [15], for $|H_1(z)H_2(z)| < 1$ for all z, $z = e^{j\omega}$ and starting from zero initial conditions, the algorithm will most probably converge to the desired solution. In the image cancellation application, $|H_1(z)H_2(z)| = |H_{dif}(e^{j\omega})H_{dif}^*(e^{-j\omega})| \leq |H_{dif}(e^{j\omega})||H_{dif}^*(e^{-j\omega})| \ll 1$, for all ω for most practical situations. All simulations carried out by the author have confirmed this behaviour.

Last, in addition to the coupled image component, the primary input y_1 could also contain additive noise n_a $(= n_I + jn_Q)$. As long as n_a is not correlated with its image n_a^*, it will not affect the performance of the signal-image separation system.

7.3.3 Real-Time Computational Load

From (7.14) and (7.15), one can find the required number of arithmetic operations as listed in Table 7.1, where $L_1 = L_2 = L$ is assumed. From the table, the required real-time computational power for the complex SAD system is given by

$$(32L)f_s \quad \text{FLOPS per second} \tag{7.23}$$

where f_s is the output sampling rate of the receiver, and FLOPS means floating point operations per second. For example, if $f_s = 1MHz$, a computational power of 64 million operations per second (MOPS) is required for a second order system ($L = 2$).

When the system has converged, then the updating of filter coefficients can be stopped and the real-time computational load reduces to $(16L)f_s$ FLOPS per second, only a half of the initial requirement. Further more, if we let $\mu_1 = \mu_2$, then the symmetric property of the adaptive filters ($w_1(k) = w_2^*(k)$) can be explored to reduce the arithmetic operations.

Table 7.1: The number of arithmetic operations of the complex SAD algorithm.

	Filtering	Coefficients updating
No. of Multiplications	$2 \times 4 \times L^\dagger$	$2 \times 4 \times L$
No. of Additions	$2 \times 4 \times L$	$2 \times 4 \times L$

† L is the order of adaptive filters w_1 and w_2.

7.4 Simulation Results

Computer simulations have been conducted in Matlab Simulink^{TM} to verify the proposed signal-image separation scheme.

7.4.1 Sinusoidal Tone Tests

Simulations with sinusoidal input tones are presented in the following paragraphs.

First, a second-order signal-image separation model was constructed. To test its frequency-dependent I/Q mismatch correction capability, a set of I/Q data composed of two complex sinusoidal tones with different I/Q mismatches was applied to the system. The input tones lie at $+0.15f_s$ and $+0.35f_s$ respectively. The first tone possesses 4.5% gain mismatch and 2.5^o phase error. The second one possesses 2.5% gain mismatch and 4.5^o phase error. The spectrum of this input is shown in Figure 7.5(a). Due to I/Q mismatches, image components can be found at $-0.15f_s$ and $-0.35f_s$ with magnitudes of -30 dB and -27.6 dB respectively. The system converges after 30000 iterations as shown in Figure 7.6(a) and (b) which plot the trajectory of the adaptive filters w_1 and w_2 (the step sizes μ_1 and μ_2 used here are 2^{-14}). From the trajectory, it can be found that $w_1(k) = w_2^*(k)$, which verifies our expectation. After processing the data, we obtained an output whose spectrum is shown in Figure 7.5(b). In the output spectrum, the two image components at $-0.15f_s$ and $-0.35f_s$ are invisible. This means that the frequency-dependent I/Q mismatches has been successfully cancelled.

Then, another set of I/Q data composed of three complex sinusoidal tones with different I/Q mismatches was applied to the second-order system. The input tones lies at $+0.15f_s$, $+0.25f_s$ and $+0.35f_s$ respectively. The first, second, and third tones possess gain mismatches of 4.5%, 3.5% and 2.5% respectively, and phase errors of 2.5^o, 3.5^o and 4.5^o respectively. The spectrum of this input is shown in Figure 7.7(a). Image components can be found at $-0.15f_s$, $-0.25f_s$ and $-0.35f_s$ with magnitude of -30 dB, -28.96 dB and -27.6 dB respectively. When processing this data, the adaptive filters $w_1(z)$ and $w_2(z)$ have their coefficient locus as shown in Figure 7.8(a) and (b) respectively. The spectrum of the output of the system is shown in Figure 7.7(b), from which one can observe that the second-order system can not completely remove the image components. However, the three image components at $-0.15f_s$, $-0.25f_s$ and $-0.35f_s$ are suppressed to -59.5 dB, -53.2 dB and -59 dB respectively. All are suppressed more than 20 dB.

A third-order system was then constructed to process the three-tone inputs of Figure 7.9(a) (same as that in Figure 7.11(a)). From the output spectrum shown in Figure 7.9(b), it is found that the third-order system eliminates completely the I/Q mismatches at those three frequencies. The coefficient locus of the third-order adaptive filters $w_1(z)$ and $w_2(z)$ when processing this data are shown in Figure 7.10.

A set of I/Q data was collected from an I/Q receiver in laboratory. This set of data has a spectrum as shown in Figure 7.11(a), and was fed to the second-order signal-image separation system. When processing this data, the adaptive filters $w_1(z)$ and $w_2(z)$ have their coefficient as shown in Figure 7.12(a) and (b) respectively (the step sizes μ_1 and μ_2 used here are 2^{-14}). The output of the system for this input has a spectrum as shown in Figure 7.11(b), from which it is found that the image component has been completely removed.

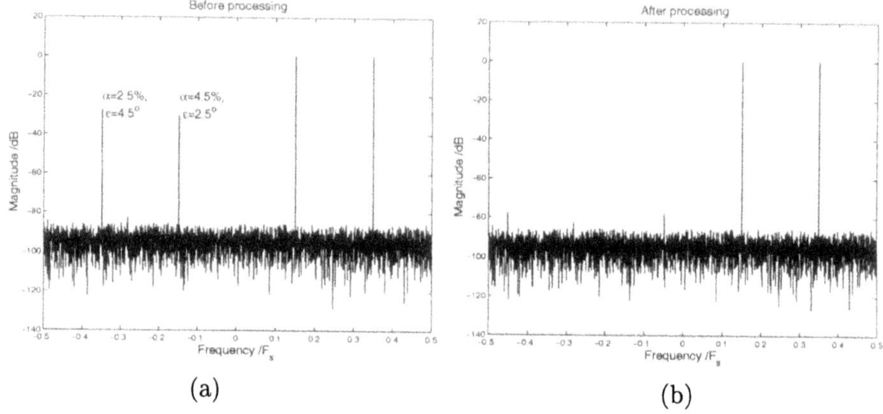

(a) (b)

Figure 7.5: Spectra in the second-order signal-image separation system: (a) two sinusoidal inputs with different gain and phase errors (α and ϵ); (b) output.

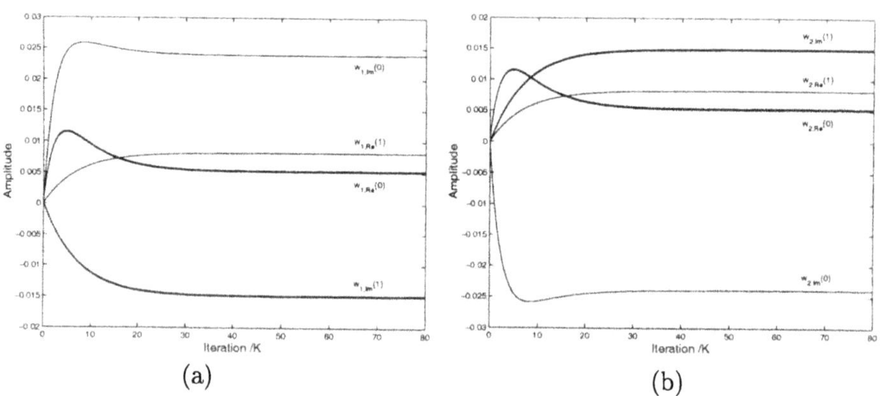

(a) (b)

Figure 7.6: Filter coefficients of the second-order SAD system responding to the two-tone input.

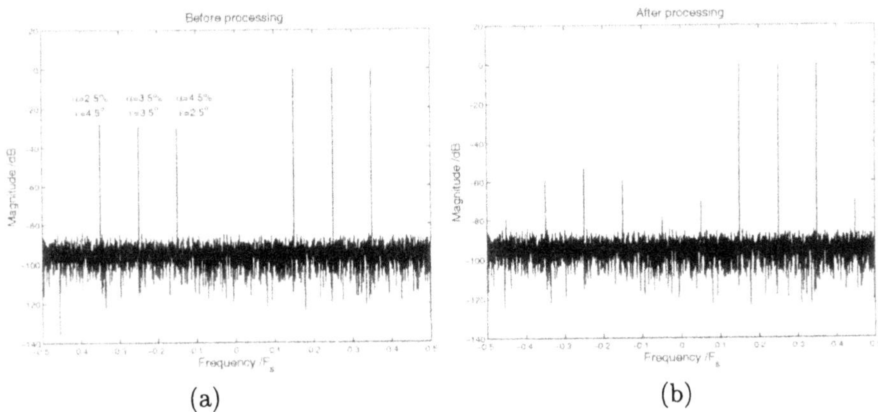

(a) (b)

Figure 7.7: Spectra in the second-order signal-image separation system: (a) three sinusoidal inputs with different gain and phase errors (α and ϵ); (b) output.

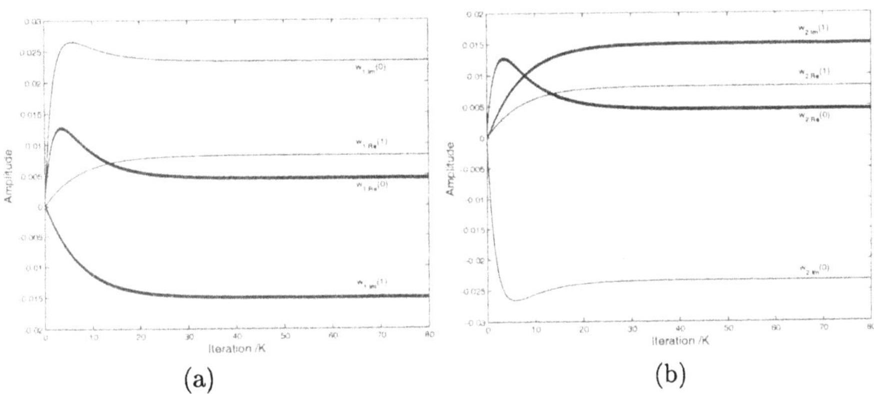

(a) (b)

Figure 7.8: Filter coefficients of the second-order SAD system responding to the two-tone input.

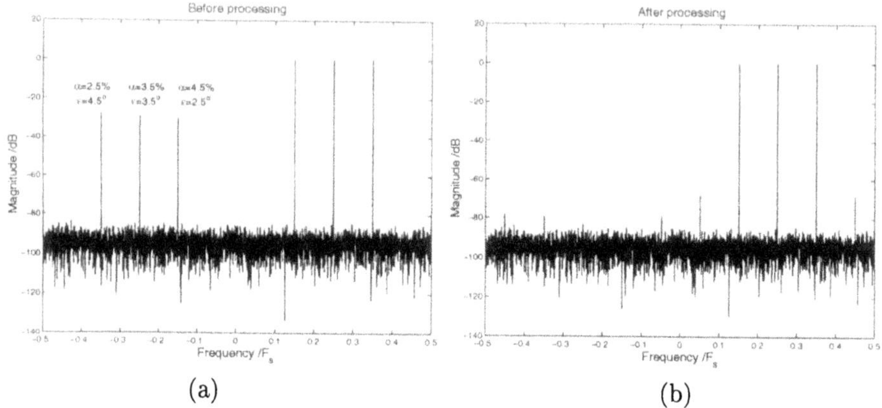

Figure 7.9: Spectra in the third-order signal-image separation system: (a) three sinusoidal inputs with different gain and phase errors (α and ϵ); (b) output.

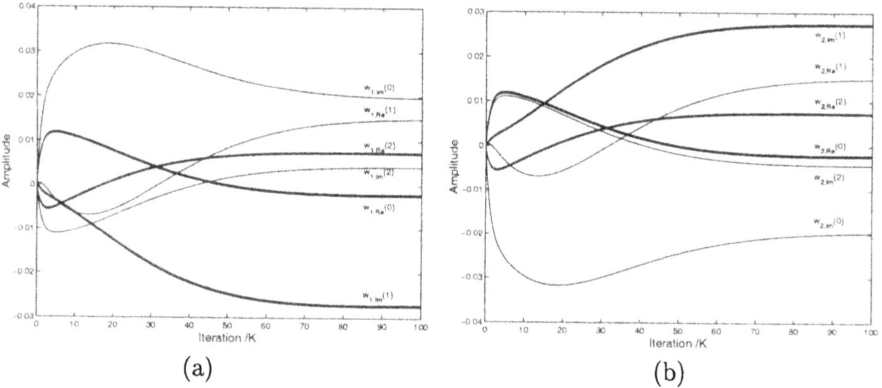

Figure 7.10: Filter coefficients of the third-order SAD system responding to the three-tone input.

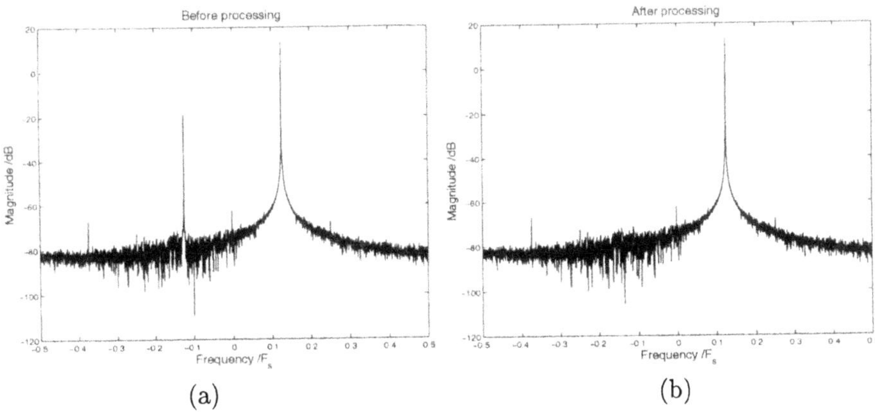

Figure 7.11: (a) Spectrum of a set of I/Q data collected from a real receiver. (b) Spectrum of the output of the second-order SAD system.

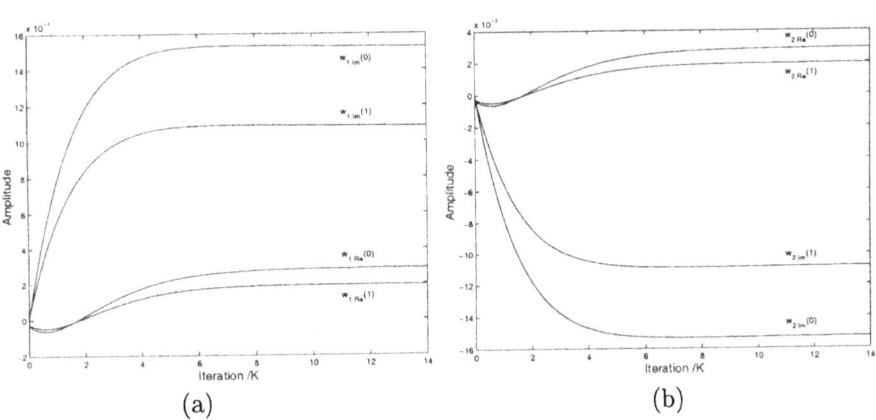

Figure 7.12: Filter coefficients of the second-order SAD system responding to the laboratory-collected I/Q data.

7.4.2 GMSK Signal Tests

To simulate the system in an environment closer to the real world, Gaussian Minimum Shift Keying (GMSK) input signals were used then.

Suppose that the receiver receives two GMSK signals. One of them is centered at $-f_s/4$ (marked as channel A) and the other at its image frequency $f_s/4$ (marked as channel B). Both of them are random GMSK signals with 3-dB bandwidth-symbol time product of 0.5. Channel B is 40 dB stronger than channel A.

Firstly, assume that the receiver has a fixed I/Q gain mismatch of 2%. The spectrum of the distorted signals is shown in Figure 7.14(a). However, the I/Q mismatch effect can not be observed in the spectrum of Figure 7.14(a) since the desired signals and the images are superposed to each other. To observe the I/Q mismatch effect, eye-pattern diagrams of channel A (the weaker signal) are plotted as shown in Figure 7.14(e) (the eye diagrams were obtained by down-converting channel A from $-f_s/4$ to DC and filtering it with a raised cosine filter). In this diagram, the eye-patterns have more than two levels at the judging instant. One can easily judge from the patterns that another signal is coupled to the desired signal. It is the coupling of channel B through I/Q mismatches in this case.

The GMSK signals are then processed by the second order signal separation system. The system converges after 5000 iterations as shown in Figure 7.14(c) and (d) which plot the trajectory of the adaptive filters w_1 and w_2 (the step sizes μ_1 and μ_2 used here are 2^{-25}). Note that only $w_{1,re}(0)$ and $w_{2,re}(0)$ are non-zeros since only frequency independent gain mismatch is considered in this case. The resulted output has a spectrum as shown in Figure 7.14(b). The eye-pattern diagrams of channel A after the processing are shown in Figure 7.14(f). Again, from the spectrum one can not find out and difference. However, the eye-patterns of Figure 7.14(f) differs very much from Figure 7.14(e). The new eye-patterns are widely opened and have negligible zero-crossing jitters. This means that the image of channel B has been successfully removed.

Now consider frequency-dependent I/Q mismatches. Assume that the receiver suffers from an frequency-dependent I/Q imbalance as shown in Figure 7.13. This mismatch condition is arbitrarily assigned, but is probably worser than most real-world conditions.

In the simulations, the mismatch is attached to the two-channel GMSK signal signal by passing its I and Q components through mismatched I and Q filters respectively. The mismatched channel filters can be constructed from their frequency domain specifications by the impulse invariance method. Since I/Q mismatches are relative, all the errors can be assigned to Q channel without

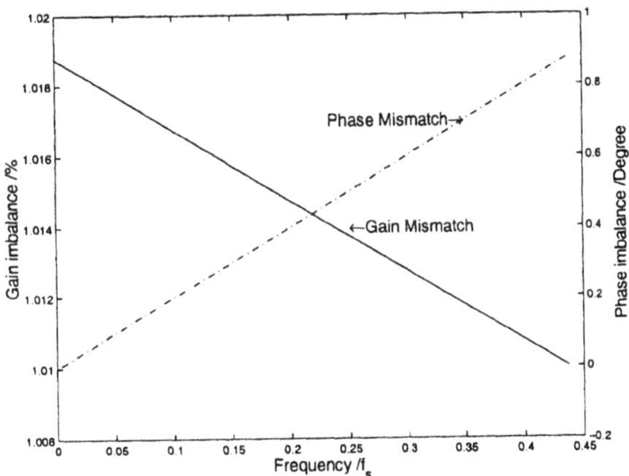

Figure 7.13: I/Q mismatches used in the simulation.

loss of generality. Let us denote the mismatched channel filters as

$$\begin{aligned} H_{ch,I}(z) &= 1, \\ H_{ch,Q}(z) &= a_0 + \sum_{i=1}^{N-1} a_i z^{-i} + j \sum_{i=1}^{N-1} b_i z^{-i}, \end{aligned} \quad (7.24)$$

where N is the order of the Q filter. For $N = 8$, the channel filter $H_{ch,Q}$ corresponding to the mismatch condition shown in Figure 7.13 has coefficients a_i, $i = 1, \cdots 7$, of $1.0137, -2.16 \times 10^{-3}, 2.20 \times 10^{-3}, -6.29 \times 10^{-4}, 7.37 \times 10^{-4}, -2.21 \times 10^{-3}, 5.84 \times 10^{-3}$ respectively, and coefficients b_i, $i = 1, \cdots 7$, of $8.93 \times 10^{-4}, -2.20 \times 10^{-3}, 1.52 \times 10^{-3}, -2.20 \times 10^{-3}, 1.78 \times 10^{-3}, -2.21 \times 10^{-3}, 2.42 \times 10^{-3}$ respectively.

The spectrum of the two-channel GMSK signal under the mismatch condition of Figure 7.13 is shown Figure 7.15(a). The eye-pattern diagrams of channel A are shown in Figure 7.15(e). The eye-patterns are very noisy and with serious zero-crossing jitters.

The two-channel GMSK signal is then processed by the second order signal separation system. The system converges after 200k iterations as shown in Figure 7.15(c) and (d) which plot the trajectory of the adaptive filters w_1 and w_2 (the step sizes μ_1 and μ_2 used here are 2^{-25}). It is also observed that $w_1(k) = w_2^*(k)$.

The resulted output has a spectrum as shown in Figure 7.15(b). Again, the frequency spectrum does not give any information. The eye-pattern diagrams

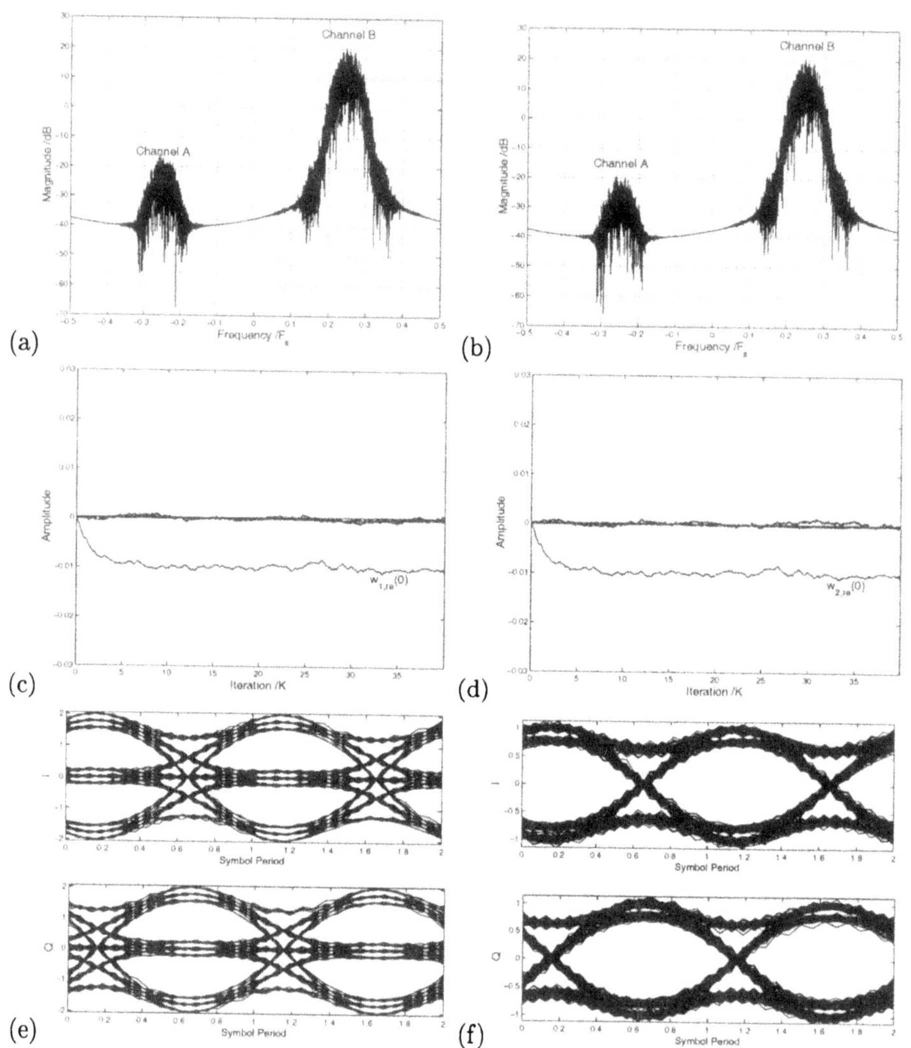

Figure 7.14: Simulation results of the second order signal separation system. (a) Input and (b) output spectrum. Coefficients of (c) filter w_1 and (d) filter w_2. Eye-pattern diagram of channel A (e) before and (f) after signal separation. 2% fixed I/Q gain mismatch is assumed.

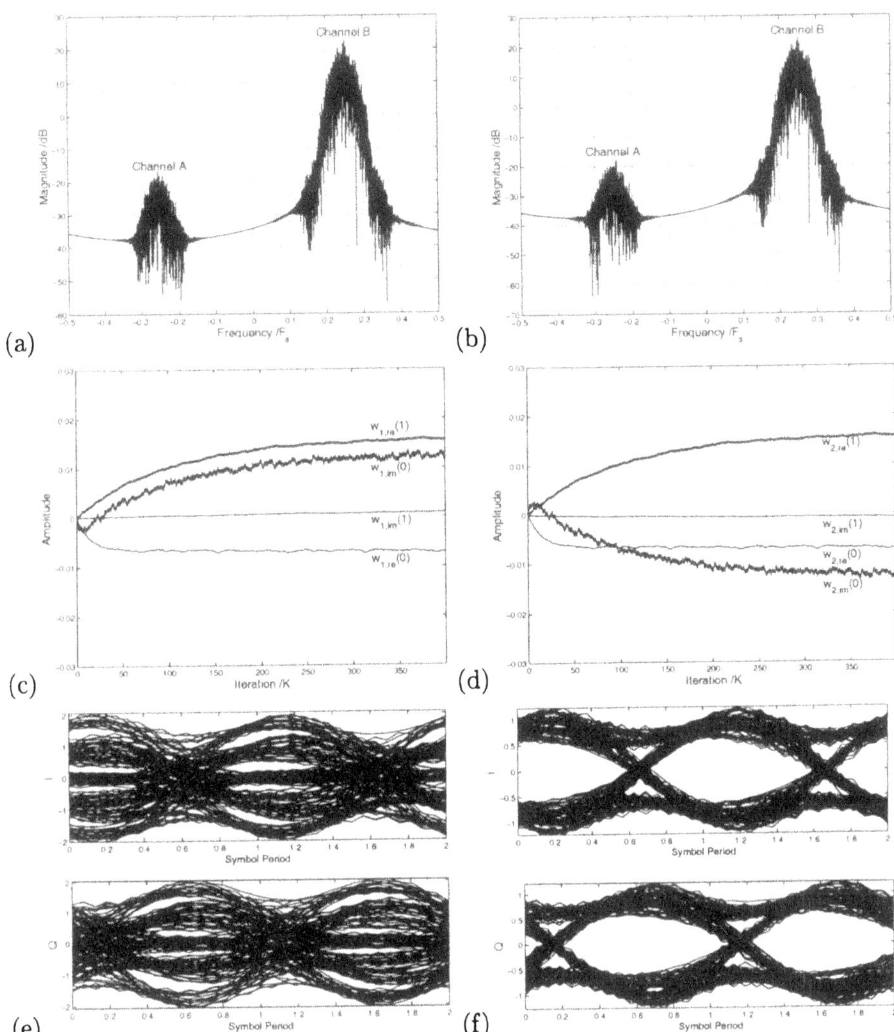

Figure 7.15: Simulation results of the second-order signal separation system. (a) Input and (b) output spectrum. Coefficients of (c) filter w_1 and (d) filter w_2. Eye-pattern diagram of channel A (e) before and (f) after signal separation. I/Q mismatches are specified in Figure 7.13.

of channel A after the processing are shown in Figure 7.15(f). We can find that the eye patterns become widely opened and only negligible zero-crossing jitter is observed. This means that the image interference due to frequency-dependent I/Q mismatches has been successfully removed by the signal separation system.

7.5 Summary

An adaptive signal separation system has been presented in this chapter for cancelling the frequency-dependent I/Q mismatches that are very important in receivers designed for multi-channel reception. This method has advantages of easy realization and the adaptation to the possible time-variation of I/Q mismatches.

Complex symmetric adaptive decorrelation scheme based on the least mean squares has been chosen. The complex conjugate of the received I/Q signal is used as the reference input of the system. The required real-time computing power is $(32L)F_s$ FLOPS per second for a system of order L. Simulations carried out in Matlab have verified the effectiveness of the proposed scheme. Cancellation of frequency-dependent I/Q mismatches has been proven.

References

[1] F.E. Churchill, G.W. Ogar, and B.J. Thompson, "The correction of I and Q errors in a coherent processor," *IEEE Transactions on Aerospace and Electronic Systems*, vol. AES-17, no. 1, pp. 131–137, Jan 1981.

[2] M.D. MacLeod, "Fast calibration of IQ digitiser systems," *Electronic Engineering, GB, Morgan-Grampian Ltd., London*, vol. 62, no. 757, pp. 41–43, 1990.

[3] M.D. Kulkarni and Alexander B. Kostinski, "A simple formula for monitoring quadrature phase error with arbitrary signals," *IEEE Tran. Geoscience and remote sensing*, vol. 33, no. 3, pp. 799–802, May 1995.

[4] J.P.Y. Lee, "Wideband I/Q demodulators: measurement technique and matching characteristics," *IEE Proc.-Radar, Sonar and Navigation*, vol. 143, no. 5, pp. 300–306, October 1996.

[5] R. A. Green, R. Anderson-Sprecher, and J. W. Pierre, "Quadrature receiver mismatch calibration," *IEEE Tran. Signal Processing*, vol. 47, no. 11, pp. 3130–3133, Nov. 1999.

[6] K.P. Pun, J.E. Franca, C.A. Leme, C.F. Chan, and C.S. Choy, "Correction of frequency-dependent i/q mismatches in quadrature receivers," *IEE Electronics Letters*, vol. 37, no. 23, Nov. 2001.

[7] Bernard Widrow, John R. Glover, John M. McCool, et al., "Adaptive noise cancelling: principles and applications," *Proceedings of IEEE*, vol. 63, no. 12, pp. 1692–1716, Dec. 1975.

[8] R. T. Compton, "The power-inversion adaptive array: concept and performance," *IEEE Trans. Aerospace Elect. Syst.*, vol. AES-15, pp. 803–814, Nov. 1979.

[9] R. L. Zinser, G. Mirchandani, and J. B. Evans, "Some experimental and theoretical results using a new adaptive filter structure for noise cancellation in the presence of crosstalk," in *Proc. Int. Acoust. Speech Signal Processing*, Mar. 1985, pp. 208–211.

[10] M. J. Al-Kindi and J. Dunlop, "Improved adaptive noise cancellation in the presence of signal leakage on the noise reference channel," *IEEE Trans. on Signal Processing*, vol. 17, no. 3, pp. 241–250, July 1989.

[11] D. Van Compernolle, "Switching adaptive filters for enhancing noisy and reverberant speech from microphone array recordings," in *Proc. Int. Acoust. Speech Signal Processing*, Apr. 1990, pp. 833–836.

[12] Y. Bar-Ness and A. Dinç, "Bootstrap: A fast blind adaptive signal separator," in *Proc. Int. Acoust. Speech Signal Processing*, Mar. 1992, vol. II, pp. 325–327.

[13] G. Mirchandani, R. L. Zinser, and J. B. Evans, "A new adaptive noise cancellatoin scheme in the presense of crosstalk," *IEEE Trans. Circuits and Systems II*, vol. 39, pp. 681–694, Oct. 1992.

[14] Ehud Weinstein, Meir Feder, and Alan V. Oppenheim, "Multi-channel signal separation by decorrelaton," *IEEE Trans. Speech and Audio Processing*, vol. 1, no. 4, pp. 405–413, Oct. 1993.

[15] S.V. Gerven and D.V. Compernolle, "Signal separation by symmetric adaptive decorrelation: stability, convergence, and uniqueness," *IEEE Trans. on Signal Processing*, vol. 43, no. 7, pp. 1602–1612, July 1995.

[16] Li Yu and W. M. Snelgrove, "A novel adaptive mismatch cancellation system for quadrature IF radio receivers," *IEEE Transactions on Circuits and Systems: - II: Analog and digital signal processing*, vol. 46, no. 6, pp. 789–801, June 1999.

[17] S.A. Jantzi et al., "The effects of mismatch in complex bandpass $\Sigma\Delta$ modulators," in *Proc. IEEE Int. Symposium on Circuits and Systems*, May 1996, pp. 227–230.

Chapter 8

Conclusions

Various circuit and system techniques for improving the image rejection performance in wideband quadrature receivers without using off-chip filters have been discussed in this book. Basically, the obstacles for high image rejection in a quadrature receiver are the quadrature errors (gain and phase imbalance between in-phase and quadrature paths) generated in the analog circuitry of a receiver. The techniques presented in this book can be categorised into two broad classes. The first class of methods is to find quadrature signal generation circuits with higher accuracy. The second class of methods is to correct or calibrate those quadrature errors generated.

One approach of precise quadrature signal generation in a wide bandwidth is to employ switched-capacitor FIR or IIR Hilbert transformers. The SC Hilbert transformers implemented in poly-phase form, which reduces the number of amplifiers required and dc offset errors, have been proposed. Moreover, the predictive correlated-double-sampling technique has been adopted to reduce the circuit's sensitivity to finite amplifier gain and bandwidth. Besides, we also discussed the pseudo-N-path implementation for the SC Hilbert transformers, which employs the fewest number of amplifiers but is more sensitive to capacitor mismatches. The proposed switched-capacitor Hilbert transformers can be used not only as the 90° phase shifter, which is critical in many quadrature receivers, but also as an image rejection filter.

Then, an application of the SC Hilbert transformers is illustrated. It was suggested to replace the traditional RC/CR phase shifter in a Hartley receiver by such an SC Hilbert transformer for improving the image rejection performance. Since the phase shifting is performed in discrete-time domain, we named this architecture as sampled-data image rejection receiver. The output signal of the receiver is inherently discrete in time. It can be directly converted to digital form by a direct IF sampling A/D converter for further processing such as demodulation. With an image rejection section before the IF digitisation can reduce the front-end anti-aliasing filtering requirement. Alternative, the discrete output of the receiver can be converted back to continuous-time domain for further processing by analog circuitry. A prototype sampled-data image rejection receiver, targeted to applications in cordless telephones, was realized in a 0.6 μm CMOS technology. The chip includes an I/Q mixer, three MOSFET-C filters with a novel frequency control circuit, an SC Hilbert transformer and an SC bandpass filter. Experiments show good results of the chip.

The second approach for precise wideband quadrature signal generation is to employ quadrature sampling circuits. If the sampling frequency is four times the signal centre frequency, then the in-phase and quadrature components of the signal can be obtained simply by alternating the sign of the sampled signal each other sample (multiplying the signal by $\{+1, 0, -1, 0, \ldots\}$ and $\{0, +1, 0, -1, 0, \ldots\}$ to obtain I and Q components respectively). Such a quadrature sampling results in the frequency down-conversion (from IF to baseband) of the signal. This kind of sampling technique can be employed in IF-sampling sigma-delta A/D converters with lowpass noise transfer functions. This approach has the advantages of IF-sampling, such as the immunity to DC offset and flicker noises, as well as the advantages of baseband noise shaping, such as reduced requirement on the noise transfer function compared to those of bandpass sigma delta modulators.

However, conventional quadrature sampling circuits suffer from the problems of capacitor mismatches and clock phase errors, which result in limited image rejection. The problem is even severer if the IF signals to be sampled are complex, such as in a low-IF receiver. This is because the image signal to be rejected is no longer the mirror of the desired signal itself as in the case of real-IF receiver, but an interferer from other radio channels which can be much stronger than the desired one. To solve the problem, a new quadrature sampling circuit with immunity to capacitor mismatches and clock phase errors has been proposed. The immunity to phase errors is achieved by employing a clocking scheme that is commonly used in switched-capacitor circuits for eliminating the signal dependent clock feed-through. The capacitor mismatch is solved by sharing some sampling capacitors among different paths. Strictly speaking, the proposed sampling circuit still has the problem of self-image generated by the mismatches. But it is much tolerable than those images from other radio channels. Actually, the proposed circuit transforms the image problem caused by I/Q mismatches to the less important self-image problem. Circuit simulations conform with the theoretical analysis.

Alternatively, an SC complex notch filter can be added in the IF sampling circuit to improve its image rejection performance. The complex notch filter is in fact formed by Hilbert transformers. Due to its analog implementations this filter also suffers from channel mismatches. Therefore, there is usually no actual advantage to add this filter. However, it was found that a first order FIR notch filter can be realized by a simple SC circuit which is insensitive to channel mismatches. The first order notch filter has a narrow bandwidth (relative to the sampling frequency), but is very effective in the case that an oversampling A/D converter are used.

The second class of methods for improving the image rejection in a quadrature receiver is to correct the I/Q imbalance generated in the analog circuitry in receivers by digital processing. This class of methods is used primarily for suppressing the so-called "self-image", which commonly exists in direct conversion receivers as well as in low-IF receivers. The self-image, however, becomes very important when one moves the channel selection function from the analog to the digital domain. This means that the receiver receives multiple channel signals at a time. Under such circumstance the "self-image" of a signal is actually an interferer from neighbouring radio channel within the receiving band.

Two new digital I/Q error compensation methods have been proposed in this book. Different from existing methods, they are wideband methods that can correct the frequency-dependent I/Q mismatches.

The first approach is a calibration method. It involves the process of test-signal injection, I/Q error estimation and I/Q error correction. The I/Q error is estimated by either a DFT method or a method based on statistical independence of the I and Q signals. The second approach is an adaptive signal-image separation method. By using the complex conjugate of the received I/Q signal as the reference, a complex symmetric adaptive decorrelation system based on least mean squares is used to suppress the self-image components. Comparing with the previous method, this method requires no test signal injections and can be adaptive to the possible time-variation of I/Q mismatches. However, it requires a higher computational power. High-level simulations have verified that both methods can largely suppress the self-image in the presence of frequency-dependent I/Q mismatches.

GPSR Compliance
The European Union's (EU) General Product Safety Regulation (GPSR) is a set of rules that requires consumer products to be safe and our obligations to ensure this.

If you have any concerns about our products, you can contact us on

ProductSafety@springernature.com

In case Publisher is established outside the EU, the EU authorized representative is:

Springer Nature Customer Service Center GmbH
Europaplatz 3
69115 Heidelberg, Germany

www.ingramcontent.com/pod-product-compliance
Ingram Content Group UK Ltd.
Pitfield, Milton Keynes, MK11 3LW, UK
UKHW020213250726

13967UKWH00003B/1437

* 9 7 8 1 4 7 5 7 3 7 3 8 7 *